FRONTIERS IN BIOORGANIC CHEMISTRY
AND MOLECULAR BIOLOGY

FRONTIERS IN BIOORGANIC CHEMISTRY AND MOLECULAR BIOLOGY

Edited by
Yu. A. Ovchinnikov
and
M. N. Kolosov

Shemyakin Institute of Bioorganic Chemistry,
USSR Academy of Sciences, Moscow

1979

Elsevier/North-Holland Biomedical Press
Amsterdam · Oxford · New York

Elsevier/North-Holland Biomedical Press 1979

All rights reserved. No part of this publication may be reproduced, stored in a retrieval system, or transmitted, in any form or by any means, electronic, mechanical, photocopying, recording, or otherwise, without the prior permission of the copyright owner

Published by:
Elsevier/North-Holland Biomedical Press
335, Jan van Galenstraat, P.O. Box 211
Amsterdam, The Netherlands

Sole distributors for the U.S.A. and Canada:
Elsevier/North-Holland Inc.
52 Vanderbilt Avenue
New York, N.Y. 10017

Library of Congress Cataloging in Publication Data

Main entry under title:

Frontiers in bioorganic chemistry and molecular biology.

"Dedicated to the memory of Mikhael M. Shemyakin, member of the USSR Academy of Sciences, pioneer of Sovjet bioorganic chemistry, in observance of the anniversary of his 70th birthday".
    Bibliography: p.
    Includes index.
    CONTENTS: Pauling, L. The nature of the bonds formed by transition metals in bioorganic compounds and other compounds.–Barton, D.H.R. The invention of organic reactions useful in bioorganic chemistry.– Woodward, R.B. Recent advances in the chemistry of natural products. [etc.]
    1.   Biological chemistry–Addresses, essays, lectures.   2.   Chemistry, Organic–Addresses, essays, lectures.   3.   Molecular biology–Addresses, essays, lectures.   4.   Shemyakin, Mikhail Mikhailovich.
I.   Ovchinnikov, IUriĭ Anatol'evich, 1934–
II.   Kolosov, M. N.   III.   Shemiakin, Mikhail Mikhaĭlovich.
QH345.F74   574.1'92   79-1188
ISBN 0-444-80072-7

Printed in the Netherlands

This volume is dedicated to the memory of
MIKHAEL M. SHEMYAKIN
Member of the USSR Academy of Sciences
Pioneer of Soviet Bioorganic Chemistry
in observance of the anniversary of
his 70th birthday

Professor M.M. Shemyakin
26.VII.1908–26.VI.1970

Preface

This book is dedicated to the memory of Mikhail Mikhailovich Shemyakin, pioneer of bioorganic chemistry in the USSR and one of the first on an international scale to realize its full impact on the other life sciences.

In his spectacular and all too short life, Shemyakin had been interested in many fields of organic chemistry, from structural theory and reaction mechanisms, to the multifaceted aspects of biologically active low molecular substances and biopolymers.

The variegated nature of the problems considered in this book are in keeping with the broad scientific interests of Mikhail Shemyakin. Preparing this volume, we, his students addressed ourselves to his friends and colleagues with the request to write for it articles on the subjects which hold their minds at the present moment. The request was favorably met by an International pleiad of scientists: Nobel laureates D.H.R. Barton (France), H.G. Khorana, L. Pauling and R.B. Woodward (USA), and V. Prelog (Switzerland); Professors A.E. Braunstein (USSR), E. Lederer (France), Yu.A. Ovchinnikov (USSR), Th. Wieland (FRG), B. Witkop (USA), and their colleagues.

We are deeply grateful to all the authors for this token of respect to Shemyakin's memory, his outstanding scientific achievements, and his significant contribution to the furthering of international scientific collaboration.

Yu.A. Ovchinnikov
M.N. Kolosov

List of contributors

D.H.R. Barton

Institute of Bioorganic Chemistry, Gif-Sur-Yvette, France

A.E. Braunstein

Institute of Molecular Biology, USSR Academy of Sciences, Moscow

E. De Clerq

Rega Institute, University of Leuven, Belgium

J. Descamps

Rega Institute, University of Leuven, Belgium

H. Faulstich

Max Planck Institute for Medical Research, Department of Organic Chemistry, Heidelberg, GFR

H. Gobind Khorana

Departments of Biology and Chemistry, Massachusetts Institute of Technology, Cambridge, Mass.

E.V. Goryachenkova

Institute of Molecular Biology, USSR Academy of Sciences, Moscow

Guang-Fu Huang

Laboratory of Chemistry, National Institute of Arthritis, Metabolism and Digestive Diseases, U.S. National Institutes of Health, Bethesda, Md.

R.A. Kazaryan

Institute of Molecular Biology, USSR Academy of Sciences, Moscow

E. Lederer

Institute of Bioorganic Chemistry, Gif-Sur-Yvette, France

Yu. A. Ovchinnikov

Shemyakin Institute of Bioorganic Chemistry, USSR Academy of Sciences, Moscow

L. Pauling

Linus Pauling Institute of Science and Medicine, Menlo Park, Calif.

L.A. Polyakova

Institute of Molecular Biology, USSR Academy of Sciences, Moscow

V. Prelog

Laboratory of Bioorganic Chemistry, E.T.H., Zürich, Switzerland

M. Robert-Géro

Institute of Bioorganic Chemistry, Gif-Sur-Yvette, France

E.A. Tolosa

Institute of Molecular Biology, USSR Academy of Sciences Moscow

P.F. Torrence

Laboratory of Chemistry, National Institute of Arthritis, Metabolism and Digestive Diseases, U.S. National Institutes of Health, Bethesda, Md.

Th. Wieland

Max Planck Institute for Medical Research, Department of Organic Chemistry, Heidelberg, GFR

B. Witkop

Laboratory of Chemistry, National Institute of Arthritis, Metabolism and Digestive Diseases, U.S. National Institutes of Health, Bethesda, Md.

R.B. Woodward

Harvard University, Cambridge, Mass.

Contents

Preface — Yu.A. Ovchinnikov and M.N. Kolosov   vii

List of Contributors   ix

Chapter 1   The nature of the bonds formed by transition metals in bioorganic compounds and other compounds — L. Pauling   1

Chapter 2   The invention of organic reactions useful in bioorganic chemistry — D.H.R. Barton   21

Chapter 3   Recent advances in the chemistry of natural products — R.B. Woodward   39

Chapter 4   Synthesis and antiviral activities of new 5-substituted pyrimidine nucleoside analogs — P.F. Torrence, E. de Clercq, J. Deschamps, Guang-Fu Huang and B. Witkop   59

Chapter 5   Structure and properties of boromycin and its degradation products — V. Prelog   87

Chapter 6   The phalloidin story — Th. Wieland and H. Faulstich   97

Chapter 7   Antiviral, antimitogenic and antimalarial activities of synthetic analogues of S-adenosyl-homocysteine — H. Robert-Géro and E. Lederer   113

Chapter 8   Ion transport in membranes and the ionophore problem — Yu.A. Ovchinnikov   129

Chapter 9   The β-replacing and α,β-eliminating pyridoxal-P-dependent lyases: enantiomeric cycloserine pseudosubstrates and the catalytic mechanism — A.E. Braunstein, E.V. Choryachenkova, R.A. Kazaryan, L.A. Polyakova and E.A. Tolosa 167

Chapter 10 Total synthesis of the biologically function gene — H. Gobind Khorana 191

Subject index 225

CHAPTER 1

The nature of the bonds formed by transition metals in bioorganic compounds and other compounds

LINUS PAULING

Organic chemistry and biochemistry deal largely with compounds of carbon, nitrogen, hydrogen, oxygen, phosphorus, and sulfur, and to some extent other elements, including the transition metals. The hemoglobin molecule, for example, contains four atoms of iron, and various enzymes important to life contain atoms of iron, manganese, cobalt, copper, zinc, molybdenum, or other transition metals. The nature of living organisms can be understood fully only when we have an understanding of the structure of the molecules of which they are composed.

The structural properties of the carbon atom are well understood. The valence electrons of carbon can be described in terms of the four orbitals that constitute the L shell of the atom. The physicists describe these orbitals in a way appropriate to field-free space as the $2s$ orbital, which is spherically symmetrical, and the three $2p$ orbitals, which have a horizontal plane of antisymmetry. The chemist, knowing that the four valence bonds formed by a carbon atom extend toward the corners of a regular tetrahedron, prefers to assign each of the four valence electrons to a tetrahedral orbital extending in the direction of the valence bond. It was pointed out in 1931 that the s orbital and the three p orbitals can be hybridized to form a set of four equivalent orbitals extending in the tetrahedral directions, and, in fact, it was shown that these tetrahedral orbitals are the best orbitals that can be formed by sp^3 hybridization [1]. Many of the properties of compounds of carbon can be discussed in a simple and straightforward way on the basis of the knowledge that the best bonds that can be formed by a carbon atom make with one another the angle 109.47°, the tetrahedral angle. For atoms such as nitrogen and oxygen, with one or more unshared electron pairs, the bond angles are slightly different, because the bond orbitals have somewhat more p character than tetrahedral orbitals.

A system of structural chemistry for compounds of the transition metals has been slowly developing over the last half century. An important part of this development has been the recognition that the elements chromium, manganese, iron, and cobalt and their congeners are often able to form nine single bonds; that is, they become enneacovalent. The factors that determine whether these transition metals are enneacovalent or have a smaller covalence are discussed in the following section. So far as I am aware, the first person to point out that the transition metals may have a large value of the covalence was Irving Langmuir, in 1921 [2]. He made use of the principle of electroneutrality that he had formulated. This principle is that the distribution of electrons in any stable compound must be such that every atom has an electric charge close to zero, the maximum deviation being one unit of electric charge [3,4]. The compound nickel tetracarbonyl, $Ni(CO)_4$, can be assigned a structure in which the nickel atom is neutral by having this atom possess one unshared pair of electrons and form a double bond with each of the four carbonyl groups. The covalence of nickel then is equal to 8. The existence of double bonds in nickel carbonyl was verified by Brockway and Cross in 1935 through their determination of the structure of the molecule by the electron-diffraction method; the nickel-carbon bond length turned out to be that for a double bond, rather than that for a single bond [5]. Structure determinations of other compounds of the transition metals have provided additional evidence of a high covalence of these metals under some circumstances.

These facts lead us to ask several questions. One of them is the following: 'Under what circumstances do atoms of transition metals have the maximum covalence 9, and under what circumstances do they have a smaller covalence?' Another question may also be asked: 'What are the favored directions for the nine covalent bonds formed by an enneacovalent transition-metal atom; that is, what are the values of the bond angles analogous to the tetrahedral angle, 109.47°, for the carbon atom?' These are the questions discussed in the following parts of this paper.

The covalence and oxidation number of the transition metals

In 1948 I published a paper with the title 'The Valences of the Transition Elements' [6]. The discussion in it was largely about the oxidation state of iron and other transition metals in the hexahydrated ions. Three general postulates were stated:

I. In a molecule, crystal, or complex ion every atom except hydrogen and the most electropositive and electronegative atoms tends to form bonds of such number and amount of ionic character as to make its residual charge very close to zero.

II. A non-bonding electron occupying an orbital may be moved into a less stable orbital without loss in energy of the system, in case that the orbital that it vacates is used for formation of a good covalent bond (the interaction energy with the central atom is the same for the bonding electron as for the non-bonding electron), whereas an unshared pair occupying a stable orbital cannot easily be moved into a less stable orbital.

III. The maximum stability for a complex in which a central atom forms several partially covalent bonds with surrounding atoms (its ligands) is obtained when the central atom has available a bond orbital for each of the ligands; but in case that the number of bond orbitals is less than the number of ligands somewhat less stable structures can also result from the resonance of the bond orbitals among the ligands.

Let us first consider an atom of cobalt. It has nine outer electrons, and also has nine orbitals available for the formation of covalent bonds, these orbitals being formed by hybridization of the s orbital, the three p orbitals, and the five d orbitals of its outer shell. It might accordingly, without electron transfer, form nine covalent bonds; for example, it might form the enneacovalent hydride, CoH_9. The hydrogen atoms would lie about 153 pm from the cobalt atom, probably at the corners of the trigonal prism with three equatorial caps. The electronegativity of enneacovalent cobalt is probably 1.7 or 1.8, and that of hydrogen is 2.1. The amount of partial ionic character of the cobalt-hydrogen bond, about 4 percent, is so small that no significant deviation from the electroneutrality rule would occur in this compound. This compound has not yet been made, but there is the possibility that it will be synthesized at some future time. An analogous complex ion, $[ReH_9]^{2-}$, has been synthetized and its structure has been determined [7]. The van der Waals radius of hydrogen is small enough to permit the formation of such a molecule or complex ion without undue crowding by the hydrogen atoms.

The van der Waals radius of fluorine is not much greater than that of hydrogen, and we might consider whether or not the compounds CoF_9, RhF_9, and IrF_9 could be synthesized. With the metal atom having electronegativity about 2 (1.8 for cobalt, 2.2 for rhodium and iridium), and fluorine having electronegativity 4, the amount of partial ionic character of the M—F bond becomes about 63 percent [8]. Each of the nine fluorine atoms would thus have an electric charge -0.63, and the atom of cobalt, rhodium, or iridium would have a charge +5.7. This deviates so greatly from the requirements of the principle of electroneutrality as to force us to conclude that the enneafluorides of these metals could not be formed. The positive charge on the metal atom could be decreased by adding five or six electrons to it. If six electrons on the cobalt atom were present in the form of unshared pairs, the atom would use three of its nine orbitals for these unshared pairs and have six available for bond formation. It might, for example, add six ammonia molecules, each held to it by a covalent bond, to form the complex ion $[Co(NH_3)_6]^{3+}$. The amount of ionic character of the cobalt-nitrogen bond is about 35 percent, so that six such bonds would reduce the charge on the cobalt atom from -3 to -1, which is compatible with the electroneutrality principle. Similarly, in the neutral complex $Co(NH_3)_3F_3$ the resultant charge on the cobalt atom is calculated to be close to zero. These complexes have zero magnetic moment, so that the structure as described, with three orbitals occupied by unshared pairs and the other six available for bonding, can be accepted as correct.

On the other hand, the complex $[CoF_6]^{3+}$ has magnetic moment corresponding to four unpaired electrons, so that the six electrons not involved in bonding would occupy five orbitals, leaving only four for the formation of covalent bonds to the six fluorine atoms, among which they resonate. The amount of charge transferred to the cobalt atom by the partial ionic character of these four covalent bonds is about 2.5 units, leaving the cobalt atom with the acceptable charge +0.5. It is accordingly the electroneutrality principle and the partial ionic character of bonds that require this ion, cobalt(III)hexafluoride ion, to have a hypoligated structure rather than a hyperligated structure.

Atoms of the transition metals are of such a size as to permit six water molecules to be ligated about them. I pointed out in 1946 that the partial ionic character of the bonds is such as to be compatible with the oxidation state +2 or +3 for them [6].

We may ask under what circumstances atoms of cobalt, iron, manganese, and chromium and their congeners can achieve enneacovalence. The enneacovalent cobalt atom is electrically neutral, and would remain electrically neutral if the ligands with which it forms nine bonds had approximately the same electronegativity, about 1.9 cobalt and 2.2 for rhodium and iridium. Atoms with this electronegativity, other than hydrogen, are so large that there might be difficulty in arranging nine of them around the central atom. The problem of steric hindrance could be overcome by having the number of ligands less than nine; that is, by forming double bonds with some atoms. Thus, cobalt can attach four carbonyl groups to it by double bonds, and have one electron and orbital available for forming an additional single bond. There are many examples of molecules of this sort, the simplest being $HCo(CO)_4$ and $(OC)_4Co-Co(CO)_4$.

The iron atom, to become enneacovalent, must have an additional electron attached to it, giving it a formal charge -1, and the atom of manganese must have two additional electrons attached and the atom of chromium three additional electrons attached to achieve enneacovalence. Electrical neutrality for the iron atom would result if the nine bonds formed by it had about 11 percent covalent character; that is, if the bonds were with atoms with electronegativity about 2.5, such as carbon. For manganese electroneutrality would result from nine bonds with atoms having electronegativity about 2.6 also, and for chromium about 2.8. Since electroneutrality needs to be achieved only to within ±1, a little leeway in these numbers is permitted, and in fact nine bonds to carbon atoms can be formed by all of these transition metals and their congeners. An example is $Cr(CO)_6$. A carbonyl group can attach itself to a transition metal atom either by a double bond, using two of the electrons of the metal atom, or by a dative single bond, in which one electron is formally transferred to the metal atom. In order that chromium become enneacovalent, three of the carbonyl groups would have to form dative single bonds with it, the other three being attached by double bonds, thus using up the nine orbitals. The six chromium-carbon bonds then have bond number 1.5.

A great many compounds have formulas and structures that are well explained by the assumption of enneacovalence of the transition-metal atoms. An example is $Os_4O_4(CO)_{12}$ [9]. The structure can be described as that of a cube with osmium and oxygen atoms on alternating corners. Since each of the oxygen atoms is forming three covalent bonds, it must have formal charge +1, having transferred an electron to the osmium atom, making it enneacovalent. Each osmium atom forms three single bonds with the adjacent oxygen atoms and three double bonds with the three adjacent carbonyl groups, thus utilizing all of its nine orbitals in bond formation.

Arrangement of bonds of enneacovalent atoms

Thousands of metallo-organic compounds have been synthesized during the last 50 years, and structure determinations have been carried out for some hundreds of them. Many investigators have striven to develop theoretical treatments of the structure of transition-metal complexes, making use almost entirely of the molecular-orbital method. These efforts to develop a quantum mechanical theory of these structures have, however, had little success, in that they have not provided any significant discussion of bond angles and bond lengths or of the relative stability of alternative structures.

I began to attack this problem many years ago [1], in part with the assistance of my students R. Hultgren [10] and V. McClure [11]. Some of the recently obtained results have been published in a series of papers [12-19]. The attack has been made entirely by use of the simple theory of hybridization [1]. In this simple theory the assumption is made that, for the transition elements of the first row, for example, the dependence of the single-electron wave functions on the radius r is sufficiently similar for the orbitals 3d, 4s, and 4p to permit attention to be focused entirely on the radial parts of these functions, that is, on their dependence on the polar angle θ and the azimuthal angle ϕ. The nine functions of interest to us are the following:

$s = 1$
$p_x = 3^{1/2} \sin\theta \cos\phi$
$p_y = 3^{1/2} \sin\theta \sin\phi$
$p_z = 3^{1/2} \cos\theta$
$d_{z^2} = (5/4)^{1/2}(3\cos^2\theta - 1)$
$d_{yz} = 15^{1/2} \sin\theta \cos\theta \cos\phi$
$d_{xz} = 15^{1/2} \sin\theta \cos\theta \sin\phi$
$d_{xy} = (15/4)^{1/2} \sin^2\theta \sin 2\phi$
$d_{x^2+y^2} = (15/4)^{1/2} \sin^2\theta \cos 2\phi$

These functions are mutually orthogonal and are normalized to 4π. The problem is to form linear combinations of them that are also mutually orthogonal and that have the maximum value in the bond direction, so as to give maximum overlap with a bond orbital of the ligand atom and thus produce the strongest bonds. The function S, equal to the value of the orbital in the bond direction, is called the bond strength of the orbital. It was shown in 1931 that a pair of equivalent sp^3 orbitals with maximum strength can be formed when their axes are in directions making the tetrahedral angle, 109.47°, with one another. It was also shown [1,10] that a pair of spd orbitals can be formed with maximum strength, 3, when their axes make either the angle 73.15° or the angle 133.62° with one another. These angles, 73.15° and 133.62°, are thus the analogues of the tetrahedral angle for the carbon atom.

Hultgren immediately attacked the problem of setting up sets of mutually orthogonal spd orbitals with the maximum strength. He developed the pertinent equations, and found that they were so complicated as not to allow their solution with the computational methods available 45 years ago. He accordingly made a simplifying assumption, that all of the orbitals had a cylindrical axis of symmetry, as has the best spd orbital. He found that sets of as many as six orbitals with strength approaching the maximum could be formed, but that the strength of additional orbitals became rapidly smaller. It was not until many years later that it was recognized that the conclusion that eight or nine good bond orbitals could not be formed by sp^3d^5 hybridization was wrong. In 1967

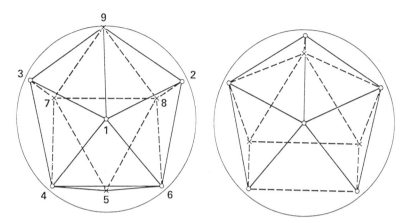

Fig. 1. Stereographic projections of directions of maximum values (bond directions) of the McClure set of nine *spd* orbitals (left), and of the set with a tetragonal axis of symmetry (right).

McClure attacked the general problem of finding the best set of nine sp^3d^5 orbitals; that is, a set of nine mutually orthogonal orbitals such that the sum of the strengths of the nine in their respective best directions would be a maximum. By developing some basic theorems and making use of a large computer, he succeeded in solving this problem [11]. He found an arrangement of nine orbitals with values of S between 2.9539 and 2.9936, oriented as shown in Fig. 1. This set probably is the best set that can be found, although there are two other sets that are almost as good.

Of the 36 bond angles of the McClure set, 21 lie between 66.49° and 92.57°, average 74.04° (only 0.89° from the best bond angle), and 15 lie between 112.92° and 140.49°, average 133.35° (0.20° from the other best bond angle).

The amount of effort involved in the discovery of this set was so great as to suggest that the direct attack made by McClure could not be effectively applied to more than a very few problems. I accordingly asked if it might not be possible to develop an approximate treatment that would give essentially the same results. It is evident that the average strength of a set of bond orbitals decreases from the maximum value 3 by greater amounts as the bond angles increase their differences from the best values 73.15° and 133.62°. The value of the strength of each of a pair of equivalent orbitals with maximum values in two directions at the angle α with one another is shown as a function of α in Fig. 2. The equation corresponding to this curve is

$$S(\alpha) = (3 - 6x + 7.5x^2)^{1/2} + (1.5 + 6x - 7.5x^2)^{1/2} \qquad [\text{I}]$$

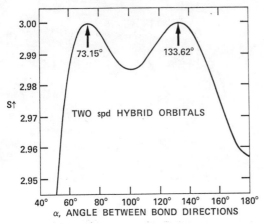

Fig. 2. The strength S in two directions at angle α of two best orthogonal *spd* orbitals as a function of α.

in which $x = \cos^2(\alpha/2)$. The straightforward derivation of this equation is given in reference [13]. The consideration of non-orthogonality of best bond orbitals with their maxima at angle α leads to another equation [14], which is, however, equivalent to Equation I:

$$S(\alpha) = 3[\{0.25 - (5\cos^2\alpha + 2\cos\alpha - 1)^2/144\}^{1/2} + 0.5]^{1/2} \qquad [\text{II}]$$

The assumption was made that the strength of a bond orbital comprising a best set, when the bonds are given a certain arrangement, can be calculated by use of the equation

$$S_i = 3 - \Sigma_j[3 - S(\alpha_{ij})] \qquad [\text{III}]$$

in which the angles α_{ij} are the bond angles for the orbital i with other orbitals j. Tests showed that this assumption has validity to within about 10 percent of the defect in strength, that is, the decrease of the strength below the value 3. For example, Racah [21] carried out a thorough quantum mechanical treatment of eight equivalent *spd* orbitals directed toward the corners of a square antiprism. He found that the best set had strength 2.9886, and that the angle of the eight bonds with the fourfold axis is 60.90°, the bond angles being 72.34° (8), 76.32° (8), 121.80° (4), and 140.93° (8), all rather close to the best bond angles. The same problem was discussed [13] by use of Equations I and III, the value of the polar angle being determined for which the bond strength is a maximum. The polar angle was found to be 61.4°, only 0.5° from the value found by Racah, and the strength S was found to be 2.9789, slightly less than Racah's value. This calculation took only a few minutes, whereas Racah's calculation must have taken weeks.

Racah also determined the best orbitals for the tetragonal dodecahedron, which is found experimentally for the ion $[Mo(CN)_8]^{4-}$. The best set of orbitals, with the sum of the strengths a maximum, consists of four with $S = 2.9954$, angle 34.55° with the fourfold axis, and four with $S = 2.9676$, angle 72.78° with this axis. The bond angles are 69.10° (2), 72.67° (4), 75.80° (8), 95.03° (4), 123.72° (2), 141.77° (4), and 145.56° (2). These bond angles agree to within one degree with the experimental values for the molybdenumoctacyanide ion. The same problem was attacked by the use of Equations I and III. This simple calculation, making the sum of the strengths a maximum, gave the polar angles 34° and 72.0°, differing by less than one degree from Racah's values. S_1 and S_2 were calculated to be 2.9944 and 2.9644, respectively, close to the values found by Racah.

A similar treatment of the McClure arrangement led to angles within about one degree of those reported by McClure, which in themselves are probably reliable to about one degree. It was then noticed that by rotating the triangle marked by points 5, 7, and 8 in the McClure arrangement (Fig. 1) through 60°, the structure that is shown at the right side of Fig. 1 is obtained. This arrangement has a fourfold axis of symmetry, passing through the circle near the top of the stereographic projection and through the center of the square face at the bottom. The arrangement can be described as a tetragonal antiprism with a cap on the upper one of the square faces. The maximum value of the sum of the values of S was found to occur for polar angles 67° and 120° for the two sets of four orbitals, the average value of S being 2.9789, as compared with 2.9816, calculated in the same way, for McClure's set. Thus, this tetragonal structure is only slightly less stable than McClure's and constitutes an alternative stable arrangement for nine sp^3d^5 bonds.

A third arrangement that is also very good, with average bond strength 2.984, is the trigonal prism with three equatorial caps. The best value of the polar angle is approximately 45°. This is the arrangement found for the nine hydrogen atoms in the enneahydridorhenium anion [7].

A test of the theory with use of observed carbonyl bond angles

A simple but convincing test of the theory of hybrid bond orbitals can be made by consideration of the OC–M–CO bond angles. Let us first consider three carbonyl groups attached to the metal atom by single bonds, with an approximate threefold axis through the $M(CO)_3$ group. This situation might arise for chromium, molybdenum, or tungsten. A carbonyl group attached by a single bond transfers an electron to the metal atom, and three such carbonyl groups would transfer three electrons, making the metal atom enneacovalent. It could then form six other bonds: for example, with the six atoms of the benzene ring in the compound benzenechromiumtricarbonyl, $C_6H_6Cr(CO)_3$. The bond orbitals for the three single bonds would be directed toward the corners of a triangular face of the coordination polyhedron. The McClure arrangement, the tetragonal antiprism with one axial cap, and the trigonal prism with three equatorial caps all have triangular faces, with some variation in the bond angles around the theoretical best value $73.15°$. In the following considerations I shall use this theoretical value, unless otherwise stated. Accordingly, we expect the bond angle $73.15°$ for the case that the bond number n is equal to one.

The bond number 1.5 occurs in chromium hexacarbonyl, $Cr(CO)_6$, in which three single bonds and three double bonds resonate among the six positions. The corresponding bond angle is $90°$. The bond angle for double bonds, $n = 2$, as in the $Co(CO)_3$ group, can be calculated by placing the axes of the double bonds midway between one of the equatorial directions and one of the basal directions

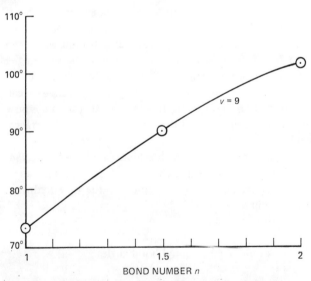

Fig. 3. The OC–M–CO bond angle in trigonal $M(CO)_3$ groups as a function of the bond number of the metal-carbon bond.

in the trigonal prism with three equatorial caps. The calculated value then is found to be 101.85°.

The three points corresponding to this calculation are shown in Fig. 3. The curve drawn through them is given by the following equation:

$$\text{Bond angle} = 73.15° + 38.7° (n - 1) - 10° (n - 1)^2 \qquad [\text{IV}]$$

Observed values of the angle OC–Co–CO in compounds containing the cobalt tricarbonyl group with approximate trigonal symmetry are given in Table 1. The average value, 101.0°, is within one degree of the theoretical value for double bonds, 101.85°. (References to the X-ray diffraction papers for the compounds listed in Table 1 and in Table 2 are given in [18].) Values of the OC–Fe–CO bond angles in Fe(CO)$_3$ group, in Table 2, average 95.6°. In all of these compounds the iron atom forms four single bonds with other carbon atoms or iron atoms, in addition to the bonds to carbonyl. The bonds to carbonyl consist of one dative single bond, which converts iron to enneacovalence by giving it an addition electron, end two double bonds. The average bond number for the iron-carbonyl bonds is accordingly 1.67. The theoretical value, calculated with use of Equation IV, is 94.5°. The agreement between the observed and calculated values is accordingly reasonably satisfactory. A similar discussion for the groups Mn(CO)$_3$ and Cr(CO)$_3$ is given in [18].

There are two reasonable arrangements for the carbonyl groups in a tetracarbonyl of a transition metal, M(CO)$_4$. One of these arrangements has an axial carbonyl and three others arranged about it in accordance with an approximate threefold axis. The second arrangement has orthorhombic symmetry, with a

Table 1. Values of OC–Co–CO bond angles in the cobalt tricarbonyl group.

Compound [R = Co(CO)$_3$]	Average angle
$C_5H_5(CO)Fe(CO)_2R$	102.5°
$C_9H_7(CO)Fe(CO)_2R$	101.8°
$C_7H_8(CO)Co(CO)_2R$	102.5°
R_3C-CR_3	98.5°
$R_3C-C_2(R_2)-C\equiv C-CR_3$	99.9°, 100.1°
$R_3C-C\equiv C-CR_3$	99.8°
$C_6F_6R_2$	100.5°
R_3CCH_3	100.5°
$R_3(CO)BH_2N(C_2H_5)_3$	103.0°
Average	101.0°

Table 2. Values of OC–Fe–CO bond angles in the iron tricarbonyl group.

Compound [R = Fe(CO)$_3$]	Average angle
$C_{10}H_{12}R_2$	93.6°
$(C_{12}H_{12}O)R_2$	95.2°
$C_7H_8R_2$	95.0°
$C_8H_{10}R_2$	95.2°
$(C_9H_{12}O)R_2$	94.2°
C_6F_8R	95.1°
CR_5	94.3°
C_4H_4R	95.5°
$(C_6H_5)_2C_2R_3$	95.0°
36 others, average	95.8°
Average for 45 compounds	95.6°

twofold axis and two vertical planes of symmetry. It can be described as involving two triangles with a shared face.

I shall first discuss the second arrangement of M(CO)$_4$ groups. There are three angles involved, with the third one dependent on the other two. We define α as the angle corresponding to the shared edge of the two triangles, and β as the four angles describing the other four edges, with γ equal to the angle between the two outer corners. For single bonds we expect α and β to have the value 73.15°; γ is then equal to 137.7°. For n = 1.5 symmetry considerations lead to $\alpha = \beta$ = 90°, γ = 180°. With the best sets of enneacovalent bond orbitals there is no way of combining them in pairs to give four double bonds in the double-triangle arrangement. I have assumed that Equation IV gives at least approximately correct values of α, β, and γ for this case. For n = 2 these are $\alpha = \beta$ = 101.9°, γ = 142°. Accordingly, Equation IV can be used for α and β for M(CO)$_4$ groups, as well as for the bond angle in trigonal M(CO)$_3$ groups.

Cobalt and its congeners have nine outer electrons and can accordingly form nine bonds without electron transfer. In their tetracarbonyls the four double bonds might lie at angles $\alpha = \beta$ = 101.9°, γ = 142°, as described above. The remaining single bond would be directed along the axis of symmetry. This arrangement of five ligands corresponds to a trigonal bipyramid distorted by drawing two of the three equatorial ligands together to the angle 101.9° and the two axial ligands toward their midpoint by 19° from the original threefold axis. An alternative way of distorting the trigonal bipyramid is to move the three equatorial ligands toward the single bonded ligand by an appropriate amount. This second structure involves much less deformation than the first, and in fact I have not found any compounds containing the cobalt tetracarbonyl group, rhodium tetracarbonyl group, or iridium tetracarbonyl group with the first arrangement of the bonds.

The theoretical value of the bond angles in HCo(CO)$_4$ with the trigonal structure can be calculated by assuming that one orbital is directed toward the hydrogen atom, five lie at the best polar angle 73.15°, and three at the other best angle 133.62°. This arrangement is approximately that found by McClure. We assign the five orbitals in the approximately equatorial belt and one of the three in the lower triangle to the three double bonds of the equatorial carbonyls, and find that the average polar angle is 83.2°. The supplement of this value, 96.8°, is the expected value of the angle between the axial carbonyl and the three equatorial carbonyl groups. In Table 3 values are given of this bond angle for nine compounds containing the tetracarbonyl group, with the cobalt atom forming a single bond with another atom in each case. The average value, 97.3°, agrees well with the theoretical value 96.8°.

For iron, ruthenium, and osmium the carbonyl groups transfer one electron to the metal atom, so that seven bonds are required for carbonyl groups, and the metal atom can form two single bonds with other groups. The arrangement of the carbonyls in the double-triangle structure is accordingly appropriate for these complexes. Observed values of α and β and their averages are given in Table 4. The average of α and β for the 11 compounds is 94.2°, with mean deviation 1.5°. Since the bond number is 1.75, the theoretical value given by Equation IV is 96.6°. There is, accordingly, reasonable agreement between the two.

In compounds containing the manganese tetracarbonyl group the manganese atom achieves enneacovalence by having two dative single bonds with carbonyl groups. The manganese-carbonyl bond number is accordingly 1.5, and the

Table 3. Values of the angle A–M–C$_{eq}$ for trigonal bipyramidal ACo(CO)$_4$ and closely related groups.

Molecule	Average bond angle
H$_3$SiCo(CO)$_4$	98.3°
F$_3$SiCo(CO)$_4$	94.6°
Cl$_3$SiCo(CO)$_4$	94.8°
In[Co(CO)$_4$]$_3$	96.7°
[(C$_2$H$_5$)$_4$N]$^+$[Br$_2$In{Co(CO)$_4$}$_2$]	95.4°
C$_6$H$_5$PAuCo(CO)$_4$	101.7°
C$_{17}$H$_{23}$As$_3$AuCo(CO)$_4$	101.9°
BrSn[Co(CO)$_4$]$_3$	93.0°
[{(C$_6$H$_5$)$_3$P}$_2$N]$^+$[HFe(CO)$_4$]$^-$	99.1°
Average	97.3°

Table 4. Values of OC—Fe—CO bond angles in the iron tetracarbonyl group.

Compound [R = Fe(CO)$_4$]	Average angles			
	α	β	αβ	γ
(C$_6$H$_5$)$_2$CCR$_2$	100.5°	92.3°	93.9°	173.0°
Cd$_4$R$_4$	97.9°	98.3°	98.2°	154.7°
C$_6$F$_4$R$_2$	97.8°	91.0°	92.4°	171.9°
C$_5$H$_5$(OC)Co(GaCl$_2$)$_2$R	97.0°	92.0°	93.0°	172.0°
(C$_6$H$_5$)$_4$C$_4$R	99.8°	93.9°	95.1°	168.1°
(CHCOOH)$_2$R	107.5°	88.7°	92.5°	176.7°
C$_8$H$_{12}$R$_2$	114.1°	90.2°	95.0°	178.3°
(CH$_3$)$_4$Sn$_3$R$_4$	93.0°	96.3°	95.6°	161.8°
(HCF$_2$CF$_2$)$_2$R	88.9°	92.7°	91.9°	171.9°
(CH$_3$)$_2$AsMn(CO)$_4$R	105.7°	90.7°	93.7°	177.0°
[(CH$_3$)$_2$AsCCF$_2$]$_2$Fe$_2$(CO)$_6$R	100.0°	93.3°	94.6°	169.6°
Averages	100.3°	92.7°	94.2°	170.5°

expected value for the bond angles α and β is 90°. The experimental values for the eight compounds listed in Table 5 lie close to this theoretical value. It may be pointed out that only for the first one the bonds formed by the manganese atom, other than the bonds to carbonyl, are well-defined single bonds. For the other compounds they may have some double-bond character, which might change the bond angles somewhat.

Table 5. Values of OC—Mn—CO bond angles in the manganese tetracarbonyl group.

Compound [R = Mn(CO)$_4$]	Average angles			
	α	β	αβ	γ
[(C$_6$H$_5$)$_2$Si]$_2$R$_2$	96.4°	89.3°	90.7°	177.7°
C$_6$H$_4$CH$_2$N(CH$_3$)$_2$R	93.1°	90.5°	91.0°	170.4°
[(CH$_3$)$_2$AsCCF$_2$]$_2$R$_2$	90.3°	93.2°	92.6°	168.7°
(CF$_3$Se)$_2$R$_2$	93.6°	88.1°	89.2°	177.4°
(C$_6$H$_5$)$_3$PCCRBr	93.4°	92.7°	92.9°	171.6°
C$_6$H$_5$NC$_6$H$_4$R	91.8°	91.7°	91.7°	172.2°
(CH$_3$)$_2$AsFe(CO)$_4$R	99.6°	89.3°	91.4°	174.0°
C$_4$F$_4$As(CH$_3$)$_2$R$_2$As(CH$_3$)$_2$	91.4°	92.6°	92.4°	172.3°
Averages [a]	93.3°	91.1°	91.6°	172.3°

[a] Not including the first compound.

The stability of octahedral complexes

For chromium and its congeners the tetracarbonyls, with three dative single bonds and one double bond, have bond number $n = 1.25$, and the metal atom might form four single bonds with other atoms. I have not found any structure determinations for substances of this sort, which seem to be unstable. There are, however, many known compounds of chromium and molybdenum in which phosphine, arsine, and thio groups are attached to the metal atom, in addition to carbonyl groups. An example of such a compound is $CHCl[CF_2As(CH_3)_2]_2\text{-}Cr(CO)_4$. All of the bond angles in this compound lie very close to $90°$, the average deviation being $1.5°$. Accordingly, the bond angles from chromium to arsenic, as well as those from chromium to carbon, have approximately the value 1.5, and we may conclude that the three single bonds and three double bonds resonate about equally among the six ligands.

A reason for the apparent stability of substances of this sort can be proposed. In a compound in which the central atom forms bonds with groups that have the capability of forming either single bonds or double bonds, the single bonds and double bonds resonate among the alternative positions, and the compound is stabilized by the energy of this resonance. The amount of resonance energy increases with the number of resonance structures. The number of resonance structures for an octahedral complex is a maximum when there are three single bonds and three double bonds: in this case there are 20 such structures. This situation arises only with the octahedral complexes of chromium, molybdenum, and tungsten. Iron, for example, can form one single bond and four double bonds, as in $Fe(CO)_5$, and the complexes stabilize to some extent by the resonance of the single bond among the five alternative positions.

Manganese, technetium, and rhenium can form pentacarbonyls in which the manganese atom also forms a single bond with another atom, as in $HMn(CO)_5$. The calculated value for the angle between the axial carbonyl group and the four equatorial carbonyl groups is $93.6°$. Many experimental values have been reported, with average $94.4°$, in reasonable agreement with the theoretical value.

In all of the above cases the observed values of carbonyl bond angles have been found to be close to those predicted by the theory of hybrid bond orbitals, and this close agreement provides strong support for the theory.

A set of enneacovalent radii for the transition metals

The problem of assigning enneacovalent radii to the transition metals was a difficult one. The difficulty arose mainly from the fact that for some of the transition metals the observed metal-metal bond lengths in different complexes are far from constant; for example, in complexes of iron the observed values fall in the range from about 245 pm to about 290 pm, without consideration of those complexes in which the iron-iron bonds were expected to be multiple bonds. In the series Cr, Mn, Fe, Co with nine valence electrons for each atom the covalent radius would be expected from general theoretical considerations and from observations on simpler structures, such as the pyrite crystal FeS_2, to decrease by a small amount from element to element, with indication that the amount of decrease is 1 pm per unit increase in atomic number. I noticed that the cobalt-cobalt distances observed in complexes of enneacovalent cobalt were nearly the same, approximately 246 pm, and concluded that the enneacovalent radius of cobalt should be taken as 123 pm [12]. With the assumption of the 1 pm change in radius to a neighboring element, I assigned radii 126 pm to Cr, 125 pm to Mn, and 124 pm to Fe, as given in Table 6. Values were assigned to the heavier transition metals in the same way (Table 6).

The explanation of the constancy of observed cobalt-cobalt single-bond lengths and the range observed for the iron-iron single bond was finally discovered [16]. In enneacovalent cobalt there is no ambiguity about the nature of the bonds; double bonds are formed with carbonyl groups, phosphine groups, and arsine groups, and single bonds are formed with other atoms. For complexes of iron, however, each iron atom must pick up an electron to achieve enneacovalence. As mentioned earlier in this paper, there is the possibility of resonance of a certain sort, between the ordinary structure $O^+\equiv C-Fe^--Fe^--C\equiv O^+$ and structures of the kind $O^+\equiv C-Fe\ Fe^-=C=O$, that is, structures in which there is a no-bond between the iron atoms, with a change of one iron-carbonyl bond from single bond to double bond. This resonance decreases the bond number for the iron-iron bond, accounting for the increase in observed bond lengths in many compounds.

These values of the enneacovalent radii account well for observed values of

Table 6. Single-bond radii for transition metals with covalence 9 (in pm).

Cr	Mn	Fe	Co	Ni
126	125	124	123	122
Mo	Tc	Ru	Rh	Pd
139	138	137	136	135
W	Re	Os	Ir	Pt
140	139	138	137	136

most of the lengths of bonds formed by the enneacovalent transition metals with other atoms. For the iron group elements the single bond to a carbon atom has length about 195 pm, and double bonds have length about 175 pm. For the bonds to phosphorus and arsine there is not very much difference between the single-bond length and the double-bond length, because there is an increase in the effective radius of the transargononic structure of phosphorus or arsenic involved in the double-bonded structures. The observed M—P and M—As bond lengths, accordingly, do not have very much value in showing what the corresponding bond number is. In some complexes, such as the cyclopentadienyl complexes of the transition metals, the metal-carbon bond length is somewhat greater than that expected for a single bond, because of resonance to structures involving carbon-carbon bonds in the ring. A detailed discussion of bond lengths in some of these compounds is given in my book The Nature of the Chemical Bond [8].

Quadruple bonds

A striking discovery was made by Kuznetzov and Koz'min [22] in the course of their X-ray investigation of the structure of a compound of rhenium. They reported that the distance between two rhenium atoms in the molecule studied by them was about 223 pm, and essentially this value was also found for several rhenium compounds by Cotton [23], who recognized the rhenium-rhenium bond to be a quadruple bond. The covalent radius of rhenium (Table 6) leads to the expected single-bond length 278 pm, and the assumption that multiple bonds have decreased bond length by the same amount as for carbon-carbon bonds would lead us to expect about 257 pm for a double bond and 244 pm for a triple bond. Thus, it is reasonable to interpret the bond length 223 pm as corresponding to a quadruple bond.

We can, however, use the simple theory of hybrid bond orbitals to predict values for a quadruple bond [12]. A good model for multiple bonds in general is that they consist of bent single bonds. For carbon compounds, for example, the double bond can be described as two single bonds between the two carbon atoms of ethylene, starting out at each atom at the tetrahedral angle $109.47°$ and bending in such a way as to converge to the other atom at the same angle. We may as a not unreasonable assumption take the curvature of the bent bonds as constant, so that each of the two bent bonds describes an arc of a circle. If the arc of the circle is assumed to have the single-bond length, 154 pm for the carbon-carbon bond, the chord is calculated to be 132 pm long, in excellent agreement with the observed length 133 pm for the carbon-carbon double bond.

Similarly, with three bent bonds described as circular arcs 154 pm long, the chord is calculated to be 118 pm, close to the observed value 121 pm. We can, accordingly, assume that a quadruple bond consists of four single bonds issuing from each atom at the appropriate bond angle, and calculate the amount of shortening in the same way [12].

For covalence 9 we may consider the arrangement of nine orbitals directed toward the corners of the tetragonal antiprism with cap. The single square face at the bottom has four bonds at the polar angle 120°, the value of the bond angle being 75.5°. With this bond angle the length of the quadruple bond is calculated to be 0.827 times the length of the corresponding single bond.

For covalence 8 of the metal atom we may use the arrangement of the tetragonal antiprism, and consider the quadruple bond to be formed with use of the four orbitals directed toward one of the square faces. The bond angle is about 76.6°, and the corresponding value of the ratio of the lengths of the quadruple bond and the single bond is 0.820. For covalence 6 the six bond orbitals may be considered to be directed toward the corners of a trigonal prism with all edges equal, with the quadruple bond formed by one of the three square sides. The value of the bond angle for this arrangement is 86.53°, and the calculated ratio of lengths of the quadruple bond and the single bond is 0.661.

There is no good arrangement of seven bond orbitals such that the coordination polyhedron has a square face. I have made a linear interpolation between the results for covalence 8 and covalence 6, and in this way obtained the value 0.741 for the ratio of lengths of the quadruple bond and the single bond.

For the chromium-chromium quadruple bond with enneacovalent or octacovalent chromium and single-bond length 252 pm, the calculated lengths for the quadruple bond are 208 and 207 pm, whereas for covalence 7 the calculated length is 187 pm and for covalence 6 the calculated length is 167 pm. A number of experimental values are known in the range around 200 pm [12,24]. The smallest quadruple-bond length for chromium reported so far is that for tetra-(dimethoxyphenyl)dichromium, 184.7 pm. In this molecule each chromium atom forms bonds with two carbon atoms of the benzene ring and two oxygen atoms of the methoxy groups, as well as the quadruple bond with the other chromium atom. The difference in the electronegativity of chromium ($x = 1.6$) and oxygen ($x = 3.5$) is such as to give the Cr—O bonds the bond number $n = 0.5$; that is, we can describe the structure by saying that chromium is heptacovalent, having received an electron from one of the two attached oxygen atoms, and that it forms a single bond with this oxygen atom and a no-bond with the other, with the single bond and the no-bond resonating to give the bond number $n = 0.5$ to each of these two bonds. For heptacovalence the calculated length of the chromium-chromium quadruple bond is 187 pm, whereas the observed length [24] is 184.7 pm, in satisfactory agreement. Similar agreement is found also for compounds of rhenium and molybdenum in which there are quadruple bonds [12]. An example of another very short quadruple bond is that for the compound tetra (2,6-dimethoxyphenyl)dimolybdenum [24]. The calculated bond length for covalence 7 is 206 pm, and the experimental value that has been reported is 206.4 pm [24], in excellent agreement.

In the ion $Re_2Br_8^{2-}$ it was observed that the four bonds to bromine atoms on one rhenium atom eclipse those on the other rhenium atom. A simple explanation of this configuration is provided by the bent-bond theory of the quadruple bond between the two rhenium atoms, in that the tetragonal antiprism describing the bonds around each of the two rhenium atoms would have one square face in common, thus causing the other two outer square faces to be in the eclipsed orientation. This explanation is completely analogous to the explanation of the planarity of the ethylene molecule given by the theory of the tetrahedral carbon atom.

The contribution of f and g orbitals to bond orbitals

In most complexes of the transition metals the number of electrons and the composition of the compound are such that a reasonable electronic structure can be assigned in which the transition-metal atoms are enneacovalent, or in some cases have a smaller valence. There are some cluster compounds, however, in which there are two extra electrons. An example is the anion $[Os_6(CO)_{15}]^{2-}$. The six osmium atoms are arranged in an octahedral complex, so that we would expect that each would form four single bonds with the four adjacent osmium atoms, with the remaining five bonds being to the attached carbonyl groups. This electronic structure would use nine orbitals for each atom of osmium, with the atoms of carbon and oxygen having their usual electronic structures. An extra orbital seems to be needed for the pair of electrons that the molecule has picked up to convert it to an anion. I have pointed out that *spd* orbitals might be expected to have some $4f$ and $5g$ character, in the same way that the *sp* orbitals of carbon have some d character [20]. The six osmium atoms have a total of 54 sp^3d^5 hybrid orbitals, so that if each had only 2 percent of f and g character there would be enough sp^3d^5 orbital liberated for one extra orbital. This phenomenon seems also to be operating in the Au_{11} cluster of gold atoms, in which the central gold atom forms 10 bonds with the surrounding 10 gold atoms [20].

Conclusion

The foregoing discussion constitutes the basis of a theory of formation of chemical bonds by the transition metals that is analogous to the simple structure theory of organic chemistry, but is more complex, in that there are two stable bond angles for the transition metals, 73.15° and 133.62°, instead of only one, the tetrahedral angle 109.47°, for the carbon atom. It is likely that this theory of hybrid bond orbitals of the transition metals can be developed in many ways similar to those in which the theory of carbon compounds has been developed, and that in the course of time it will lead to a better understanding of such phenomena as the catalytic activity of transition-metal catalysts, which have great importance in practical chemistry, the catalytic activity of metal-containing enzymes, and in general the nature of biochemical reactions involving compounds of transition metals.

References

1. Pauling L. (1931) J. Amer. Chem. Soc., *53*, 1367–1400.
2. Langmuir I. (1921) Science, *54*, 59–67.
3. Langmuir I. (1920) Science, *51*, 605.
4. Pauling L. (1948) J. Chem. Soc., 1461.
5. Brockway L.O. and Cross P.C. (1935) J. Chem. Phys., *3*, 828–833.
6. Pauling L. (1948) In the Victor Henri commemorative volume: *Contribution à l'Étude de la Structure Moléculaire*, Desoer, Liège.
7. Abrahams S.C., Ginsberg A.P., and Knox K. (1964) Inorg. Chem., *3*, 558–567.
8. Pauling L. (1960) *The Nature of the Chemical Bond*, Cornell University Press, Ithaca, New York.
9. Bright D. (1970) Chem. Commun., 1169–1170.
10. Hultgren R. (1932) Phys. Rev., *40*, 891–907.
11. McClure V. (1970) *Ph.D. Dissertation*, University of California, San Diego.
12. Pauling L. (1975) Proc. Natl. Acad. Sci. USA, *72*, 3799–3801.
13. Pauling L. (1977) Proc. Natl. Acad. Sci. USA, *72*, 4200–4202.
14. Pauling L. (1976) Proc. Natl. Acad. Sci. USA, *73*, 274–275.
15. Pauling L. (1976) Proc. Natl. Acad. Sci. USA, *73*, 1403–1405.
16. Pauling L. (1976) Proc. Natl. Acad. Sci. USA, *73*, 4290–4293.
17. Pauling L. (1978) Acta Crystallograph., *B34*, 746–752.
18. Pauling L. (1978) Proc. Natl. Acad. Sci. USA, *75*, 12–15.
19. Pauling L. (1978) Proc. Natl. Acad. Sci. USA, *75*, 569–572.
20. Pauling L. (1977) Proc. Natl. Acad. Sci. USA, *74*, 5235–5238.
21. Racah G. (1943) J. Chem. Phys., *11*, 214.
22. Kuznetsov B.G. and Koz'min, P.A. (1963) Zh. Strukt. Khim. *4*, 55–62.
23. Cotton F.A. (1975) Chem. Soc. Rev., *4*, 27–53.
24. Cotton F.A. (Dec. 12, 1977) Chem. & Eng. News, p. 34.

Yu.A. Ovchinnikov and M.N. Kolosov (eds.) Frontiers in Bioorganic Chemistry
and Molecular Biology © 1979, Elsevier/North-Holland Biomedical Press

CHAPTER 2

The invention of organic reactions useful in bioorganic chemistry

DEREK H.R. BARTON

Summary

The invention of new organic reactions has been classified in terms of the processes of conception, misconception and accident. Although most of the organic reactions used in synthesis have been discovered by accident, it is possible sometimes to invent by conception. These statements are illustrated by analysis of past work.

It is an interesting philosophical exercise to consider how new organic reactions of use in synthesis are discovered. I have argued for some time that this process of discovery or invention can be classified under the three headings of conception, misconception and accident. It is the purpose of this article to analyse some of my own work in these terms. In making this article a personal one, I do not wish to suggest in any way that others have not been more fruitful in the invention of new reactions.

In 1954 it was desirable, in order to confirm the stereochemistry of euphol in terms of that of lanosterol, to invert the C_{10} methyl group of lanosterol acetate (part formula (1)). Since we knew how to transform (1) into the dienone (2), it seemed reasonable that the weakened C_1-C_{10} bond of (2) would permit equilibration at C_{10} by, say, photolysis (see (3)). In collaboration with Dr. Edward Wheeler, we showed that dienone (2) was rapidly transformed into a new compound by irradiation, but this compound was not the isomer (3), it was an umbellulone whose structure was fully determined later [2] to be (4).

These observations reminded us of the rich, but uninterpreted, photochemistry of that classical natural product santonin (5). Since santonin could be pur-

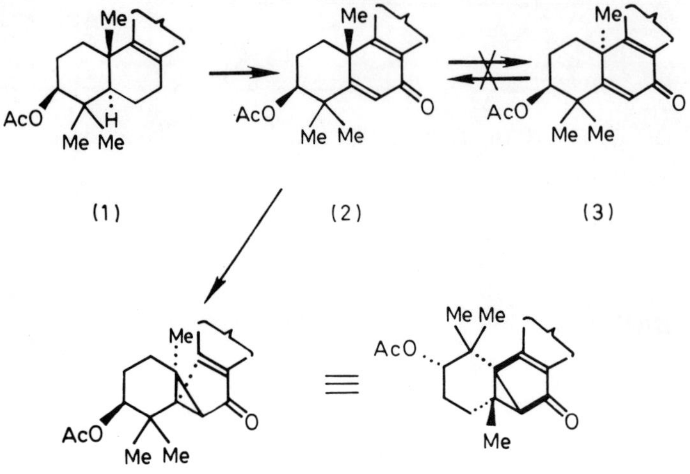

(1) (2) (3)

(4)

chased, it was an easier starting material than the dienone (2). Starting in 1955 we quickly determined [3] the structure of the classical compound isophotosantonic acid lactone (6), which is formed by photolysis in aqueous acetic acid. Photolysis in neutral medium gave a new photoisomer of santonin, lumisantonin. This compound was also an umbellulone, and on treatment with hot aqueous acetic acid it also gave compound (6). We elucidated [4] the structure (7) of

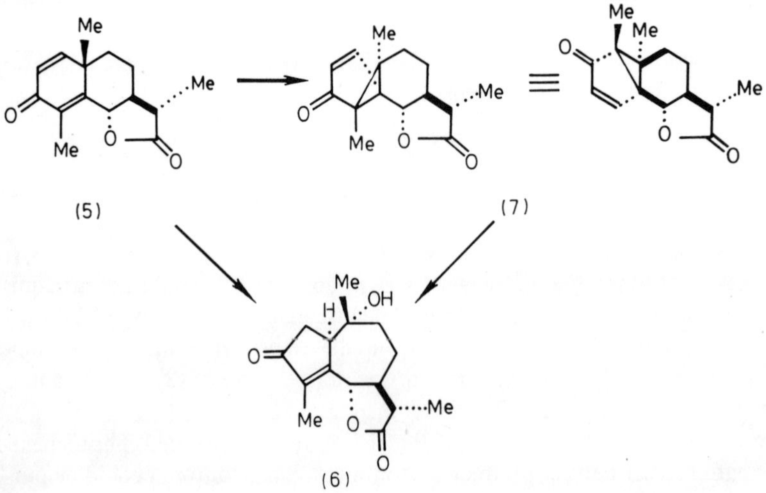

(5) (7)

(6)

lumisantonin at the same time as other colleagues [5]. It is my opinion that the remarkable and unique molecular acrobatics undergone by santonin in the presence of light had a major influence in rekindling interest in organic photochemistry. It is amusing that our role in the research began with a *misconception* of what the effect of light should be on a cross-conjugated cyclohexadienone.

Following the work on santonin, it did not require any intellectual effort to conceive that linear conjugated dienones of type (8) should also show interesting photochemical behavior. Photolysis of dienones of type (8) produced [6] a photostationary equilibrium with diene-ketens of type (9). In the presence of the appropriate nucleophiles (NuH) these ketens were quantitatively captured to give the derivatives (10). This is one of best organic photochemical reactions discovered from the point of view of high quantum and chemical yield. It has been much studied by Prof. G. Quinkert of Frankfurt and his work has recently culminated [7] in an elegant and practical synthesis of crocetin (11), with the photochemical cleavage of the *bis*-dienone (12) as the critical synthetic step. However, I think that the original discovery [6] of this reaction must be classified as *accident,* not as conception.

When our Research Institute for Medicine and Chemistry (RIMAC) opened its

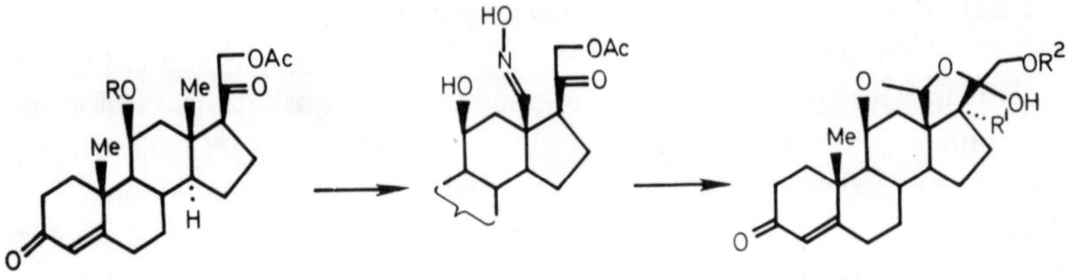

(15) R=H
(16) R=NO

(17)

(13) R¹= OH, R²=H
(14) R¹= R²=H
(18) R¹=H, R²=Ac

doors in Cambridge, Mass. in 1958 the first project proposed by the Director, Dr. M.M. Pechet, was the synthesis of 17α-hydroxyaldosterone (13). My reaction to this was that it would be better to synthesise first the important hormone aldosterone (14) which was then available in only mg amounts. After a moment of reflection I wrote down what seemed to me an ideal process viz. the conversion of corticosterone acetate (15) into its nitrite (16) and the rearrangement of the latter into the 18-oxime (17) of aldosterone acetate. This rearrangement was to be effected by a new photochemical reaction as conceived in Scheme 1.

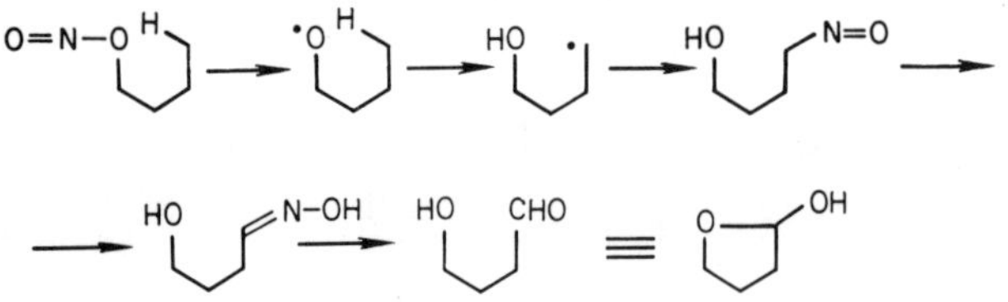

Scheme 1.

Treatment of the 18-oxime with nitrous acid would then furnish aldosterone acetate (18), the biological equivalent of aldosterone (14). This proposal worked well [8], and just over a year later we had 60 g of pure aldosterone acetate at a time when the supply in the rest of the world was in mg.

By the same general scheme 17α-hydroxyaldosterone (13) was synthesised, as well as many analogues [9]. Improved aldosterone syntheses with respect to cost have also been reported by us recently [10], again all being based on nitrite photolysis for the functionalisation of C_{18}. I consider this a good example of the invention of a useful new reaction by *conception*.

The next new reaction which I would like to describe is a process for converting a saturated acid into a γ-lactone. Such a process is important in the synthesis and partial synthesis of sesquiterpenoid lactones. At first [11] we explored the feasibility of the process in Scheme 2. An acid (19) might be converted to its radical (20) which might rearrange with hydrogen abstraction to the isomeric radical (21), which, by electron transfer oxidation, might furnish eventually the desired γ-lactone (22). To reduce this scheme to practise, we treated [*11*] carboxylic acids with lead tetra-acetate and iodine [*12*] under irradiation with a tungsten lamp at room temperature. We obtained excellent yields of the corresponding decarboxylated iodide (23). Clearly (20) had decarboxylated at once to radical (24). Although this decarboxylation process was discovered by *misconception*, it remains a useful organic reaction in synthesis because of the mild conditions involved.

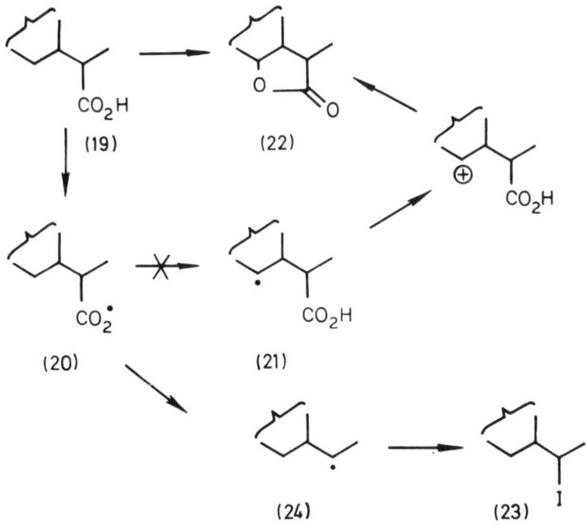

Scheme 2.

Frustrated by the easy decarboxylation of radical (20), we considered what derivative of an acid would afford radicals which would permit the desired γ-functionalisation. Clearly an amide (Scheme 3) would be suitable. We conceived that an amide (25), on treatment with lead tetra-acetate iodine would afford an diiodo-amide (26) which on photolysis would furnish the nitrogen radical (27). Lacking a facile equivalent of decarboxylation this radical should rearrange to radical (28), which would be captured by iodine to give the intermediate (29). The latter, being a γ-iodo-amide, should cyclise easily to the imino-ether (30), which by mild reduction and acid hydrolysis would afford the desired γ-lactone (22). It was possible to reduce all this to practise and to show the intermediacy of imino-ethers of type (30) [*13*]. In addition we were able to replace the toxic lead tetra-acetate by a new reagent t-butyl hypoiodite, which

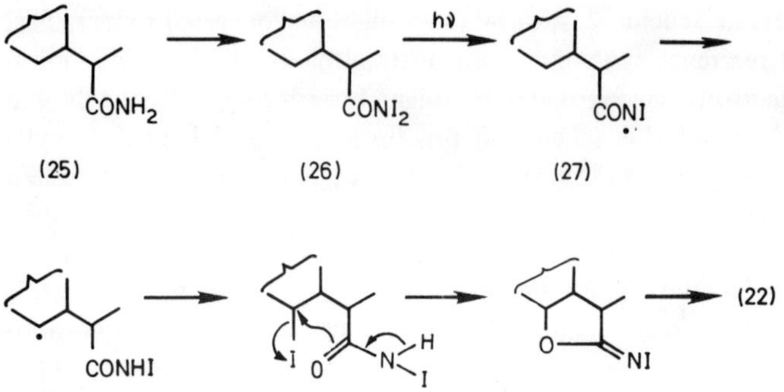

Scheme 3. (structures 25–30 leading to 22)

has powerful positive iodination properties. This γ-lactone synthesis is another good example of invention by *conception*. We used the reaction later [*14*] to convert oestrone (31) into 18-hydroxyoestrone (32), a compound found in such minute amounts in female urine that its biological activity (like that of aldosterone) could only be evaluated properly after being made available by synthesis.

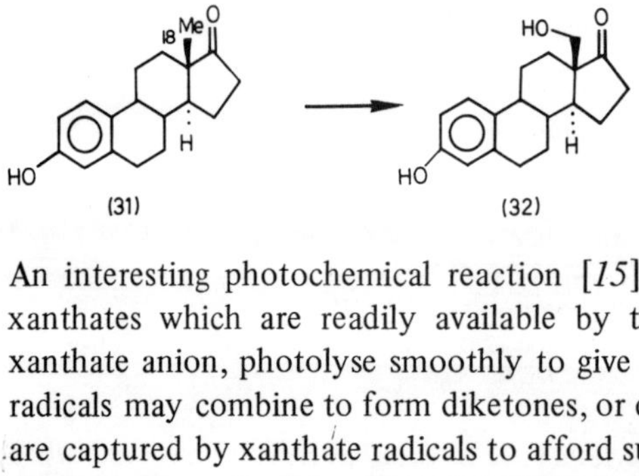

(structures 31 → 32)

An interesting photochemical reaction [*15*] is summarised in Scheme 4. Acyl xanthates which are readily available by the reaction of acyl chlorides with xanthate anion, photolyse smoothly to give acyl and xanthate radicals. The acyl radicals may combine to form diketones, or decarbonylate to give radicals which are captured by xanthate radicals to afford substituted xanthates. Xanthate radicals can also, of course, dimerise. This reaction was discovered by *conception* thinking of which bond in an excited triplet acyl xanthate would most likely give rise to radicals. It is a reaction which should receive further study and utilisation.

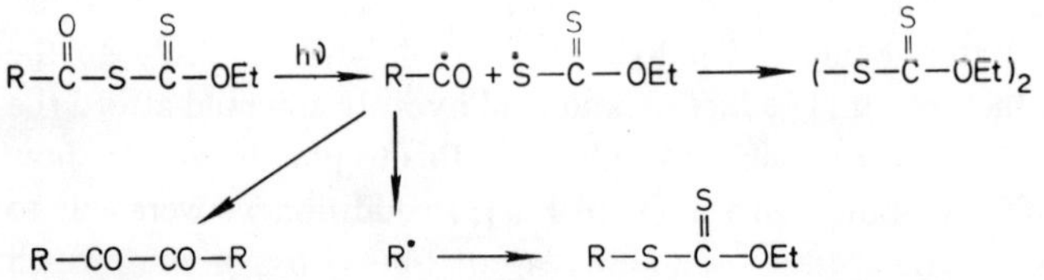

$$R-\overset{O}{\underset{\|}{C}}-S-\overset{S}{\underset{\|}{C}}-OEt \xrightarrow{h\nu} R-\overset{\cdot}{C}O + \overset{\cdot}{S}-\overset{S}{\underset{\|}{C}}-OEt \longrightarrow (-S-\overset{S}{\underset{\|}{C}}-OEt)_2$$

$$R-CO-CO-R \quad\quad R^\cdot \longrightarrow R-S-\overset{S}{\underset{\|}{C}}-OEt$$

Scheme 4.

Another photochemical reaction of wide generality was discovered entirely by *accident*. In connection with a study of α,ω-rearrangements [16] it was necessary to investigate the synthesis of thionobenzoates like the cholesterol derivative (33). By chance a yellow solution of this thiobenzoate was left on the bench exposed to laboratory lightning for several days. I noted that the yellow colour faded. When the solution was finally worked up it contained a quantitative yield of 3,5-cholestadiene (34) and of thiobenzoic acid, conveniently isolated (air oxidation) as the disulphide (35). We subsequently [17] showed that this reaction is of wide generality (Scheme 5), but requires the presence of a functional group X (double bond, aryl, oxygen, nitrogen etc.) which can stabilise the intermediate carbon radical (36).

So far all the new reactions discussed have been photochemical. In reality, no synthetic chemist will use a photochemical reaction when he can carry out the same transformation thermally in an equivalent yield. We will, therefore, now turn to thermal reactions and make the same type of analysis of their originality.

During work [18] on the key limonoid limonin (37) we had occasion to treat this natural product with potassium t-butoxide in t-butanol at room temperature under nitrogen gas which had not been adequately deoxygenated. A chromophore was slowly produced which was ultimately shown to be due to the diosphenol (38). The latter played an important role in the determination of the structure of limonin.

This autoxidation reaction, which we discovered only by *accident*, involves the reaction of the anion of the ketone with oxygen as indicated in the formulae (39) and (40). Of course, this type of reaction was already known in the literature, but it was not appreciated that good yields of single products could be obtained. The diosphenol (38) was, in fact, obtained in excellent yield. We subsequently showed [19] that under controlled conditions good yields could often be obtained in ketonic anion oxidation and that, in particular, the hydroxylation of steroids at C_{17}, an essential step in the synthesis of adrenocortical steroids, could be effected efficiently. With some significant modifications this reaction has become important [20] in the synthesis of 16β-methylcorticoid hormone analogues.

(37) (39) (40) (38)

In the course of the structural work [18] on limonin we had need of a reaction which would convert an aldehyde group, corresponding to the C_{10}-hydroxylmethyl group of limonin (see (37)), into a methyl group under mild conditions at room temperature. Working with neopentyl aldehyde (41) we conceived [21] that the hydrazone (42), on oxidation with iodine in the presence of a mild base (triethylamine) would furnish the diiodide (43) which, of course, would be smoothly reduced to neopentane (44). In fact the diiodide (43) was not known, but it was easily prepared in reasonable yield by the oxidation reaction, an example then of *conception*. Scheme 6 shows the proposed mechanism of this reaction, which has subsequently been widely studied by Sternhell and his colleagues [22]. In suitable cases it provides a synthesis of vinyl iodides instead of geminal diiodides. In the absence of the mild organic base only the (synthetically uninteresting) azine is formed.

A long-standing problem in the synthesis of corticosteroids is the conversion of unsaturated ketones of the type (45) into dienones of type (46). We thought

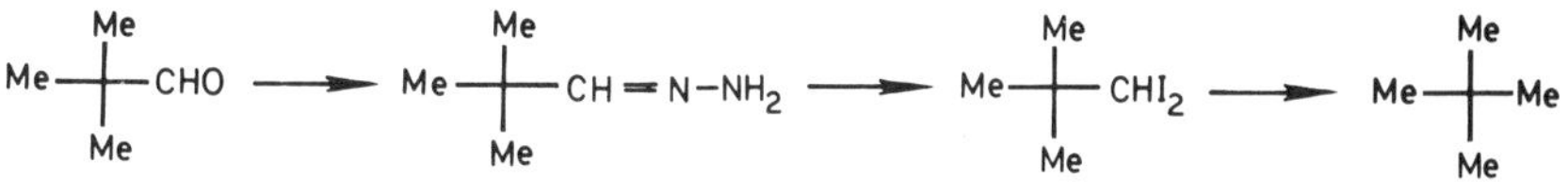

Scheme 6.

(41) Me₃C-CHO → (42) Me₃C-CH=N-NH₂ → (43) Me₃C-CH₂ → (44) Me₃C-Me

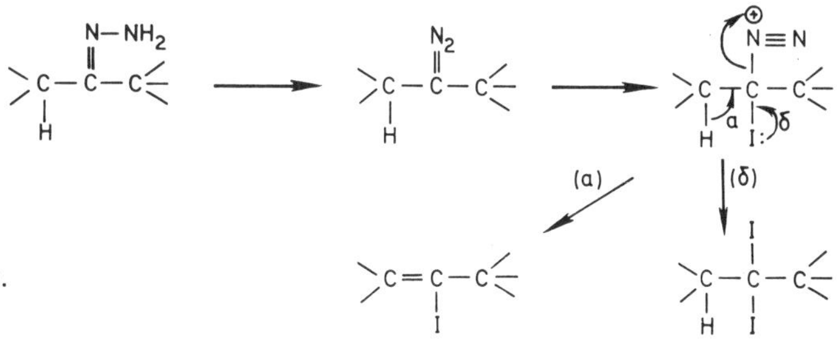

$$\text{>C-C(=N-NH}_2\text{)-C<} \xrightarrow{} \text{>C-C(N}_2\text{)-C<} \xrightarrow{} \text{>C-C-C<}$$

(a) → >C=C-C<
(δ) → >C-C-C<

that the dehydration of ketonic nitrones of the type (47) might serve this purpose (Scheme 7). The nitrones (47) are as readily prepared as oximes. On treatment with tosyl chloride in pyridine at room temperature the product, formed in good yield, was not the imine (48) but the amide (49). Further work showed that the yield is increased by working with moist pyridine and that the other

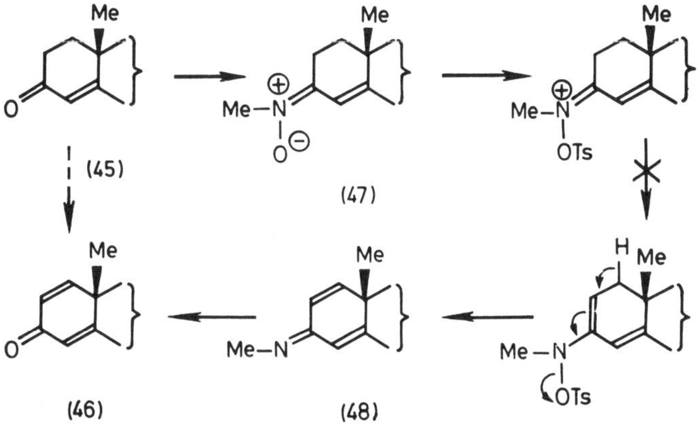

Scheme 7.

(45) → (47) → [Me-N⁺-OTs] ⤬

(46) ← (48) ← [Me-N, OTs]

nitrone stereoisomer (50) gives exactly the same amide (49). The process is a good example of a reaction discovered by *misconception* [23]. It provides a convenient alternative to the Beckmann rearrangement, but is different in that the amide produced is always formed by the movement of the more electron-rich bond, independently of geometry. The mechanism given in (51) summarises the known facts.

(47) [Scheme structures 50, 51, 49 with TsCl, Py, H₂O]

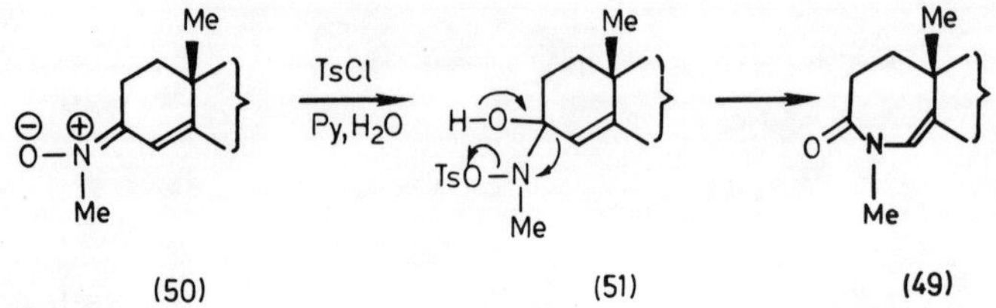

A well-known problem in the synthesis of corticosteroids is the conversion of a 20-ketone of type (52) into the corresponding dihydroxyacetone of type (53). We conceived a new solution to this problem which is summarised in Scheme 8. The idea was that a 20-ketoxime of type (54) on vigorous acetylation in the presence of a mild base (pyridine) would afford a vinylic N,O-diacetate which would spontaneously rearrange. In fact, the product of the reaction, formed in excellent yield, was the vinylic diamide (55), hydrolysed under mild conditions to the amide (56). However, this reaction, discovered by *misconception*, provides an equally good solution to the problem [24]. Acetoxylation of (56) with

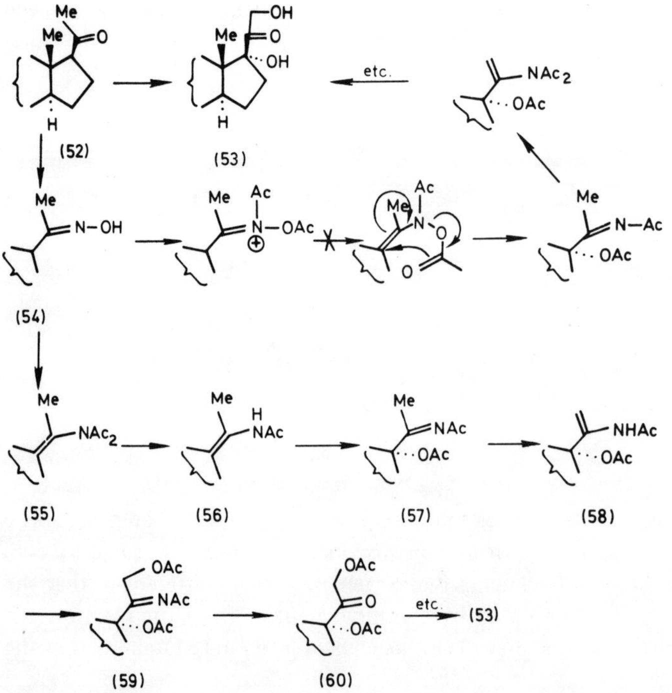

Scheme 8.

lead tetra-acetate in pyridine gave the acetoxy-derivative (57). This isomerised with anhydrous acetic acid quantitatively to the vinyl isomer (58). Further acetoxylation gave the imine (59) which on mild aqueous acid hydrolysis afforded the desired side chain diacetate (60). The reactions from (56) onwards can, in principle, be carried out in 'one pot'.

During the course of this work an interesting new rearrangement was discovered by *accident* [25]. When 17-ketoximes of type (61) were heated under reflux with pyridine-acetic anhydride a diacetylenamide was formed in excellent yield. It was, however, not the expected product of reaction (62), but the C_{13}-epimer (63). Acid hydrolysis gave smoothly the known ketone (64) which had previously been synthesised rather inefficiently by photochemical isomerisation of the normal 17-ketone (65).

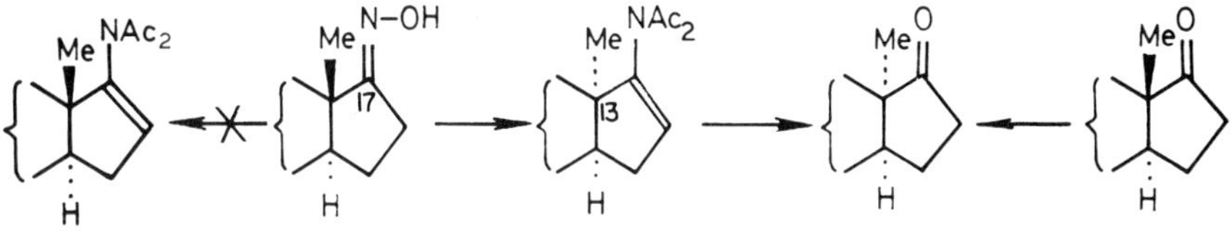

(62) (61) (63) (64) (65)

The proposed mechanism for these reductive acetylation reactions is summarised in Scheme 9. The key intermediate is the vinyl radical (66). This type of radical explains the epimerisation at C_{13} (see Scheme 9). The alternative inter-

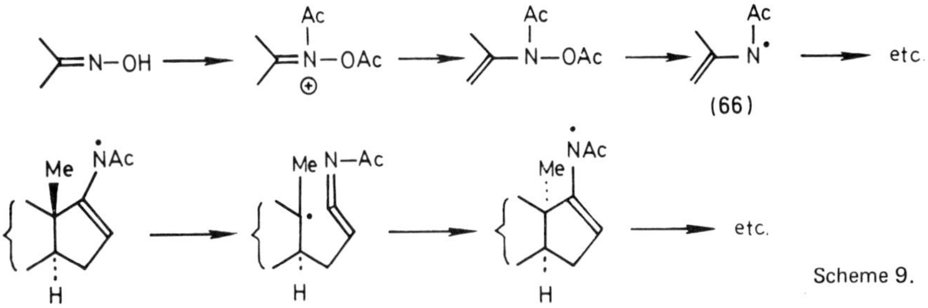

Scheme 9.

mediate, a cation (see (67)), would, in the presence of pyridine, be irreversibly deprotonated to (68) or equivalent ring opened product.

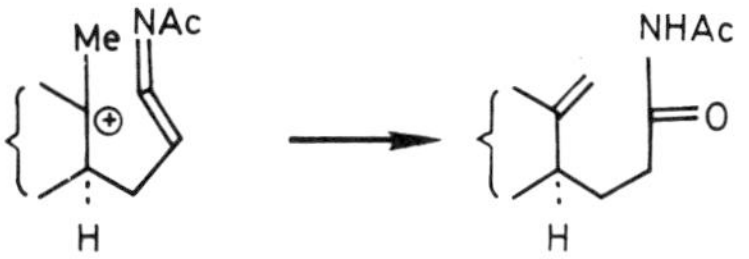

(67) (68)

We turn now to some problems of carbohydrate chemistry. In connection with the modification of aminoglycoside antibiotics it was desirable to invent a reaction for the replacement of secondary hydroxyl by hydrogen without using an ionic (SN_2 or SN_1) reaction. The SN_2 process is particularly liable to steric hindrance. In contrast to ionic reactions, radical processes are much less liable to steric restraints and to undesired cationic rearrangements. It was decided, therefore, to design a radical reaction. Scheme 10 shows the solution to the problem

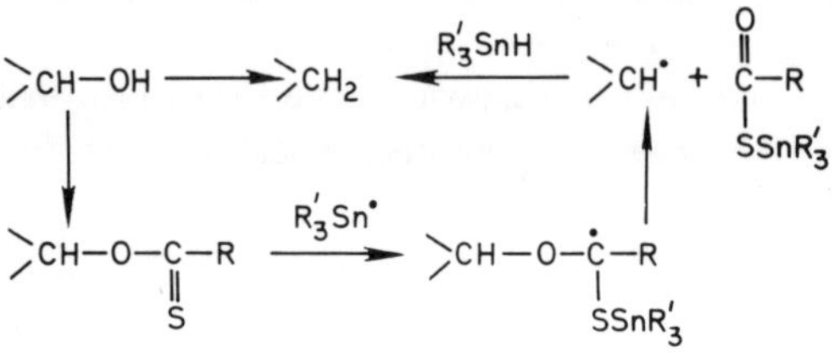

$$>CH-OH \longrightarrow >CH_2 \xleftarrow{R'_3SnH} >CH^{\bullet} + \overset{O}{\underset{SSnR'_3}{\overset{\|}{C}-R}}$$

$$>CH-O-\underset{S}{\overset{\|}{C}}-R \xrightarrow{R'_3Sn^{\bullet}} >CH-O-\underset{SSnR'_3}{\overset{\bullet}{C}-R}$$

(R=Ph,H,Me,SMe etc.)

Scheme 10.

[26], and represents a good example of the discovery of a reaction by *conception*. In practise, the most practical derivative for the reduction is the methyl xanthate, a derivative which can be always easily and cheaply prepared in excellent yield. A good example is the conversion of the 3-xanthate (69), prepared in excellent yield from diacetoneglucofuranoside, to 3-deoxy-glucose via the deoxy-compound (70). This is a reaction which has been done in good yield on a kilogramme scale.

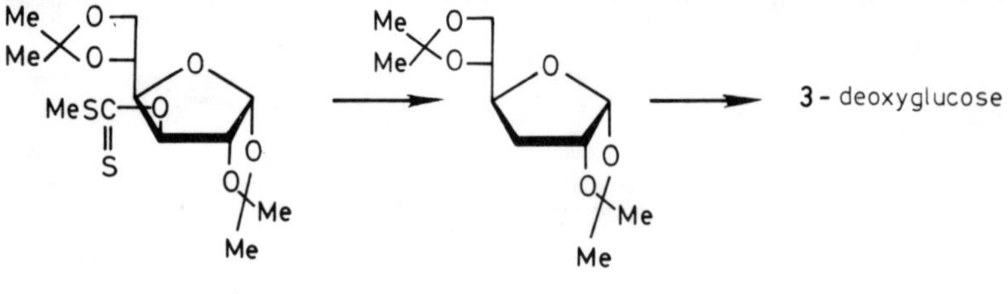

(69) ⟶ (70) ⟶ 3-deoxyglucose

It is interesting to pursue the synthesis of polydeoxy-sugars by radical processes. The first consideration is what happens when a derivative of a 1,2-glycol is reduced by a tin hydride. Two of the possible reactions are summarised in Scheme 11. Route (a) was observed [26] as the major pathway when the dithione benzoate (71) was reduced the product being a mixture of isomers (see (72)). The much more interesting reaction in route (b) was, however, observed [27] for the dixanthate (73) which afforded the olefin (74) in good yield. This

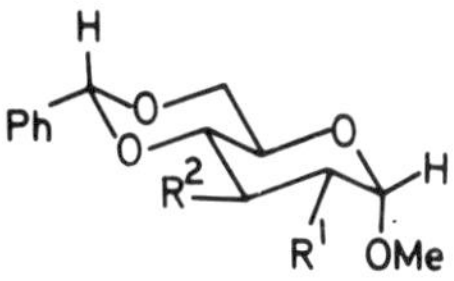

(71) $R^1=R^2=O-\overset{S}{\underset{\|}{C}}-Ph$

(72)

(74)

(73) $R^1=R^2=O-\underset{\underset{S}{\|}}{C}-SMe$

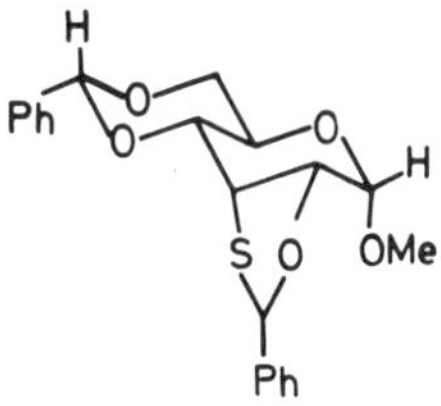

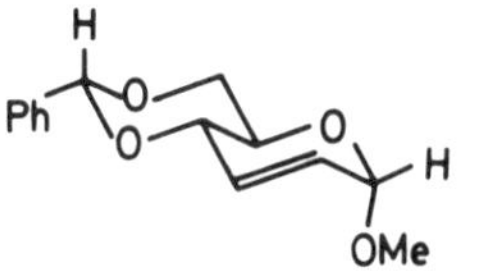

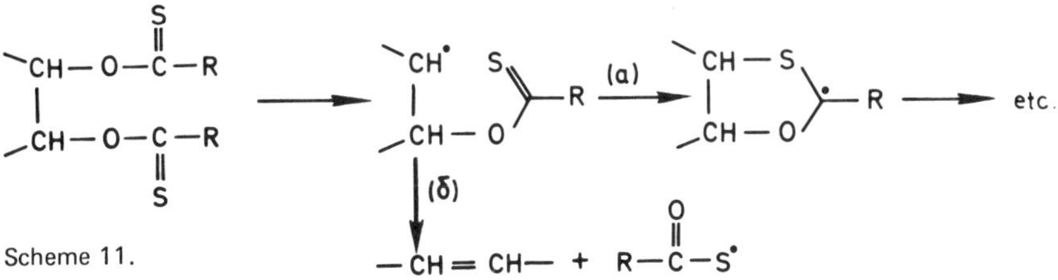

Scheme 11.

reaction of 1,2-dixanthates has been generalised [27]. Because of the mild conditions involved, it makes an excellent addition to the methods available for the conversion of 1,2-glycols to olefins. It can be classified as a reaction of *conception*.

When a primary-secondary thione carbonate such as (75) was alkylated [28] with methyl iodide it afforded in high yield, by an SN_2 ring opening of the cationic intermediate, a primary iodide (76). By conventional reactions this gave the corresponding 6-deoxy-sugar. The regiospecificity of this reaction is controlled by the SN_2 mechanism (attack on primary more facile than attack on secondary carbon). It occurred to us [29] that a radical ring opening of the thione carbonate (75) should proceed with the inverse regiospecificity (secondary more stable than primary carbon radical). This idea proved to be correct. Tributyl tin hydride reduction of (75) gave as the only product (after working up) the 5-deoxy-sugar (77). We regard this reaction as an obvious extension of what preceded it (see above).

Another problem in sugar chemistry that we were able to solve was the design of a new ether synthesis different from the traditional synthesis of Williamson. The latter involves alkylation of the anion of the appropriate alcohol under strongly basic conditions and is relatively unselective. We wanted to design a syn-

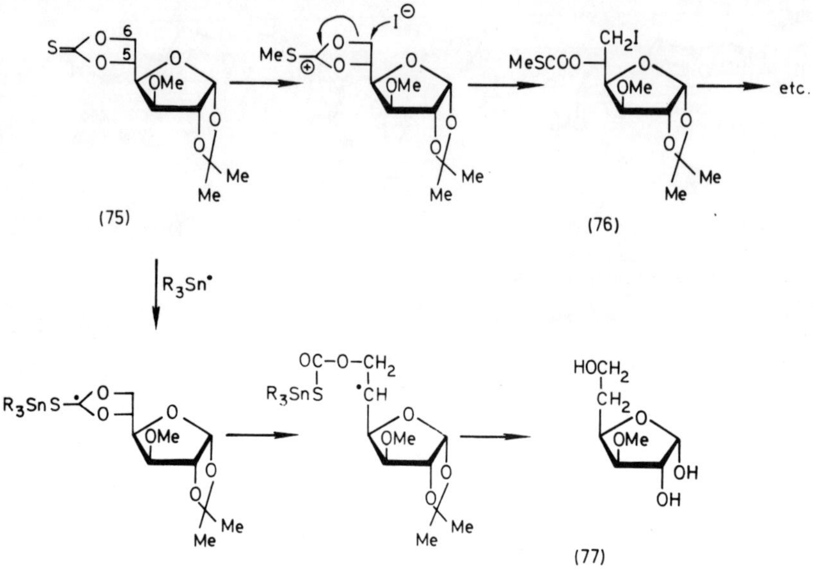

[Scheme showing structures (75), (76), (77) with tin hydride reactions]

thesis which would work under neutral conditions and with the same selectivity as can be developed in an acylation reaction.

During our work on the selective deoxygenation of secondary alcohols [26] we developed a new, efficient and general method for the synthesis of thione esters. This is summarised in Scheme 12. This method depends on the reaction of an alcohol with a Vijlsmeir reagent to give an intermediate of type (78). Normally the dimethylamino-Vijlsmeir reagent is used but a reagent from a more bulky secondary amine can be employed if greater steric selectivity is needed. Reaction of intermediates of type (78) with pyridine-H_2S gave excellent yields of thiono-esters. We naturally examined the reaction of the intermediates of type (78) with NaHSe and thus obtained a wide range of seleno-esters of type (79). Reduction with Raney nickel gave the required ethers in high yield [30].

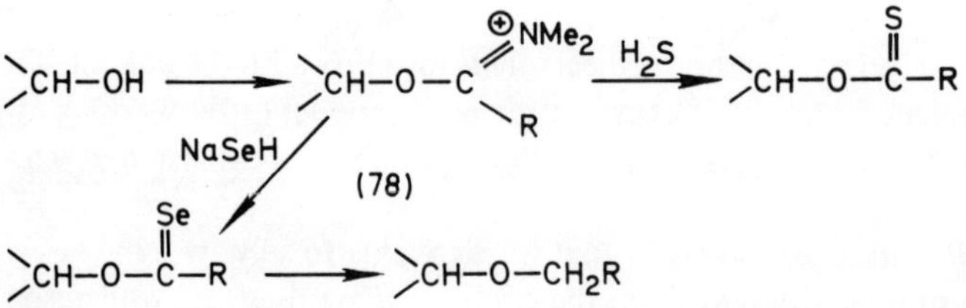

$R = H, Me, Et, CH_2Ph$ etc.

Scheme 12.

Although this ether synthesis should be useful, I feel that its conception is too obvious to justify classification.

Finally a reaction which was discovered by accident. Some ten years ago now we conceived that hypofluorites of the type R_fOF should be good 'positive' fluorine reagents. Much work at RIMAC since that time has confirmed this *conception*. A typical example is the synthesis [31] of compounds of the general type RNF_2. In the course of this work we considered also potential syntheses of compounds of the type R-NHF. Scheme 13 shows one possible route to such

$$R-NHCOCF_3 \xrightarrow{F^\oplus} R-\underset{|}{\overset{F}{N}}-COCF_3 \xrightarrow{?} RNHF \qquad \text{Scheme 13.}$$

compounds in which a trifluoroacetamide might be N-fluorinated with (say) CF_3OF and then hydrolysed under very mild basic conditions to give RNHF.

In fact, when the trifluoroacetyl derivative (80) was treated with CF_3OF or with F_2 it gave in good yield the monofluoro-derivative (81). We consider [32] that both CF_3OF and F_2 are reacting by a process which inserts F^+ into that C-H bond which best supports a positive cleavage (a tertiary C-H bond). In steroids

(80) → $\xrightarrow[\text{or } F_2]{CF_3OF}$ → (81)

this new fluorination mechanism shows remarkable regio-selectivity. For example, cholesterol acetate dichloride (82) affords the 17α-fluoro-derivative (83). The pregnenolone derivative (84) gives only 14-fluorination. After further processing 14-fluoroprogesterone (85) can be obtained.

(82) $\xrightarrow{F_2}$ (83)

(84) → CF₃OF etc. → (85)

References

1. Barton D.H.R., Wheeler E.L. (1954) unpublished observations.
2. Barton D.H.R., McGhie J.F., Rosenberger M. (1961) J. Chem. Soc., 1215.
3. Barton D.H.R., de Mayo P., Shafiq M. (1957) J. Chem. Soc., 929.
4. Barton D.H.R., de Mayo P., Shafiq M. (1957) Proc. Chem. Soc., 205; (1958) J. Chem. Soc., 140.
5. Arigoni D., Bosshard H., Bruderer H., Büchi G., Jeger O., Krebaum L.J. (1957) Helv. Chim. Acta 40, 1732; see also Cocker W., Crowley K., Edward J.T., McMurry T.B.H., Stuart E.R. (1957) J. Chem. Soc., 3416.
6. Barton D.H.R., Quinkert G. (1958) Proc. Chem. Soc., 197; (1960) J. Chem. Soc., 1.
7. Schnieder K.R., Dürmer G., Hache K., Stegk A., Quinkert G., Barton D.H.R. (1977) Chem. Ber., 110, 3582.
8. Barton D.H.R., Beaton J.M. (1960) J. Amer. Chem. Soc., 82, 2641; (1961) 83, 4083.
9. Akhtar M., Barton D.H.R., Beaton J.M., Hortmann, A.G. (1963) J. Amer. Chem. Soc., 83, 1512.
10. Barton D.H.R., Basu N.J., Day M.J., Hesse R.H., Pechet M.M., Starratt A.N. (1975) J. Chem. Soc. (Perkin Trans. 1), 2243.
11. Barton D.H.R., Serebryakov E.P. (1962) Proc. Chem. Soc., 309; Barton D.H.R., Faro H.P., Serebryakov E.P., Woolsey N.F. (1965) J. Chem. Soc., 2438.
12. cf. Heusler K., Kalvoda J. (1964) Angew. Chem. Int. Ed. Engl., 3, 525.
13. Barton D.H.R., Beckwith A.L.J. (1963) Proc. Chem. Soc., 335; Barton D.H.R., Beckwith A.L.J., Goosen A. (1965) J. Chem. Soc., 181.
14. Baldwin J.E., Barton D.H.R., Dainis I., Pereira J.L.C. (1968) J. Chem. Soc. C, 2283.
15. Barton D.H.R., George M.V., Tomoeda M. (1962) J. Chem. Soc., 1967.
16. Barton D.H.R., Prabhakar S. (1974) J. Chem. Soc. (Perkin Trans. 1), 781.
17. Barton D.H.R., Bolton M., Magnus P.D., West P.J. (1973) J. Chem. Soc. (Perkin Trans. 1), 1580; and earlier work there cited.
18. Arigoni D., Barton D.H.R., Corey E.J., Jeger, O. (1960) Experientia, 16, 41; Barton D.H.R., Pradhan S.K., Sternhell S., Templeton J.F. (1961) J. Chem. Soc., 255.

19. Bailey E.J., Elks J., Barton D.H.R. (1960) Proc. Chem. Soc., 214; Bailey E.J., Barton D.H.R., Elks J., Templeton J.F. (1962) J. Chem. Soc., 1578.
20. Gardner J.N., Carlton F.E., Gnoj O. (1968) J. Org. Chem., *33*, 3294.
21. Barton D.H.R., O'Brien R.E., Sternhell S. (1962) J. Chem. Soc., 470.
22. Campbell J.R., Pross A., Sternhell S. (1971) Aust. J. Chem., *24*, 1425; and earlier papers.
23. Barton D.H.R., Day M.J., Hesse R.H., Pechet M.M. (1971) Chem. Commun., 945; (1975) J. Chem. Soc. (Perkin Trans. 1), 1764.
24. Boar R.B., McGhie J.F., Robinson M., Barton D.H.R., Horwell D.C., Stick R.V. (1975) J. Chem. Soc. (Perkin Trans. 1), 1237; Boar R.B., McGhie J.F., Robinson M., Barton D.H.R. (1975) ibid., 1242.
25. Boar R.B., Jetuah F.K., McGhie J.F., Robinson M.S., Barton D.H.R. (1975) J. Chem. Soc. Chem. Commun., 748; (1977) J. Chem. Soc. (Parkin Trans. 1), 2163.
26. Barton D.H.R., McCombie S.W. (1975) J. Chem. Soc. (Perkin Trans. 1), 1574.
27. Barrett A.G.M., Barton D.H.R., Bielski R., McCombie S.W. (1977) J. Chem. Soc. Chem. Commun., 866.
28. Barton D.H.R., Stick R.V. (1975) J. Chem. Soc. (Perkin Trans. 1), 1773.
29. Barton D.H.R., Subramanian R. (1976) J. Chem. Soc. Chem. Commun., 867; (1977) J. Chem. Soc. (Perkin Trans. 1), 1718.
30. Barton D.H.R., Hansen P.E., Picker K. (1977) J. Chem. Soc. (Perkin Trans. 1), 1723.
31. Barton D.H.R., Hesse R.H., Pechet M.M., Toh H.T. (1974) J. Chem. Soc. (Perkin Trans. 1), 732.
32. Barton D.H.R., Hesse R.H., Markwell R.E., Pechet M.M., Toh H.T. (1976) J. Amer. Chem. Soc., *98*, 3034; Barton D.H.R., Hesse R.H., Markwell R.E., Pechet M.M., Rozen S. (1976) ibid., *98*, 3036.

Yu.A. Ovchinnikov and M.N. Kolosov (eds.) Frontiers in Bioorganic Chemistry
and Molecular Biology © 1979, Elsevier/North-Holland Biomedical Press

CHAPTER 3

Recent advances in the chemistry of natural products

R.B. WOODWARD

For some time, we have been interested in the development of methods for the synthesis of macrolides, a large class of naturally-occurring substances, many of which exhibit significant antimicrobial or antifungal activity.

Among the characteristic structural features of the macrolides are large lactone rings, in whose main carbon chains numerous asymmetric carbon atoms, variously substituted, are situated. Consequently, the entire ensemble presents a stereochemical pattern of peculiar intricacy — one which, as we shall see shortly, presents a particularly interesting and exciting challenge for the synthetic chemist.

Although our investigation has been carried out on a very broad basis, especially in its early stages, we have chosen erythromycin (Fig. 1 — I; the Roman

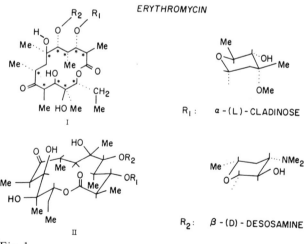

Fig. 1.

numerals denominate a particular structure, depicted in the figures) as a special objective. Erythromycin is among the best known of the macrolides, in part because it has achieved a notable position in medical practice through its effectiveness in combatting numerous pathogenic micro-organisms. Structurally, it is a prototypical macrolide, containing a fourteen-membered lactone ring, adorned by no less than ten nuclear asymmetric carbon atoms.

The armoury of the synthetic chemist is richly endowed with weapons of great effectiveness in the stereoselective, or often stereospecific, generation of newly created asymmetric centers *within rigid systems*, as best exemplified by cyclic – and especially fused polycyclic – systems. By contrast, the construction, in a desired stereochemical sense, of asymmetric arrays in flexible, open-chain systems, is a relatively little-known art, in which stereoselectivity is rare, or little understood when observed, and generalisations are dangerous. Into which class do macrocyclic assemblages such as those found in the macrolides fall? One's initial instinct is to discern a parallel with the flexible, open-chain systems. And by and large, we feel that that premonition is correct. To be sure, when dealing with a particular macrolide, and substances derived from it by relatively small changes, some degree of molecular rigidity has been observed. For example, erythromycin itself adopts the conformation rather crudely depicted in Fig. 1 –II in the crystalline state, as shown by X-ray crystallographic studies, and considerable evidence is available, especially from proton magnetic resonance studies, that a not dissimilar conformation is maintained in solution by the antibiotic, and by numerous acyl derivatives. Furthermore, scrutiny of II reveals three- to five-carbon chain segments which are very nearly perfectly staggered, and whose substituents occupy positions in space very reminiscent of the axial and equatorial dispositions of substituents attached to rigid six-membered rings. Nevertheless, we felt it would be unwisely sanguine to presume that such conformational preferences would survive deep-seated constitutional changes. Nor would any such adventitious regularities as might be found be applicable without modification to macrocyclic arrays of different size. None of these considerations should be so construed as to deny the possibility that a codification of stereochemical generalisations applicable to saturated macrocyclic systems will not one day be achieved. But we chose to adopt the position that we were facing the high challenge of building, stereoselectively, linear systems containing many asymmetric centers, disposed with respect to one another in a sense preordained by the given structure of our ultimate objective.

It is now appropriate to point out that the stereochemical relationships – both relative and absolute – along the main chain of erythromycin, while of course embodied in the formulae I and II (Fig. 1), are in many respects most succinctly represented through use of the Fischer convention, as in Fig. 2 – III;

ERYTHROMYCIN
FISCHER CONVENTION
III

R·I·P·
IV

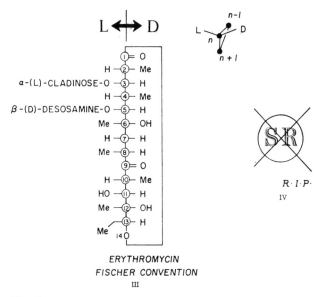

Fig. 2.

we shall use this representation, or appropriate portions of it, to monitor our progress in elaborating the desired chain. Parenthetically, it may be noted that the Cahn–Ingold–Prelog S/R convention, whatever its merits in other connections, is highly confusing when used in discussions of closely related compounds within the area of our interest, and should be strictly avoided (cf. Fig. 2 – IV).

In sum, the considerations presented in the foregoing analysis led, in the earliest stages of our planning, to the basic decision set forth in Fig. 3. It is gratify-

DECISION Nr. I:

BE NEITHER AWED NOR TEMPTED
BY THE 14-MEMBERED LACTONE RING

THUS

1. CONSTRUCT AN APPOSITELY SUBSTITUTED 13-HYDROXYTRIDECANOIC ACID

2. CONVERT THE HYDROXY ACID TO THE CORRESPONDING LACTONE

Fig. 3.

ing to note that our early presumption that it should be possible to bring about cyclisation of appropriately substituted long-chain hydroxy acids to the desired macrocyclic lactones has been shown to have been well founded, through brilliant work by other groups — notably, the synthesis of methymycin, a twelve-membered lactone, by Masamune, and of erythronolide B, a close relative of the nuclear moiety of erythromycin, by Corey.

How now did we propose to attack the problem we had delineated through our first decision? Our general approach is shown in Fig. 4.

Strong words, these! Are they justifiable? What might they mean in detail? Let us now inspect Fig. 5. In V (Fig. 5) we see a *cis-dithiadecalin system* — a bicyclic array of two fused six-membered rings — variously ornamented with substituents at asymmetric carbon atoms. Those substituents have been so chosen, and so oriented, as to be redolent of the substitution pattern in the C.9 through C.13 portion of the erythromycin main chain. Such a bicyclic system should be possessed of considerable rigidity; indeed it should be able to adopt only one or the other of the two conformations depicted in VI and VII (Fig. 5). Further, should one of the latter be preferred, that preference would be a consequence of the substitution pattern — e.g. were C.9 to be part of a very large grouping of atoms, VII, with C.9 equatorially oriented, would be favoured. And now, our key presumption: we supposed that in constructing and operating upon bicyclic dithiadecalin systems, we should be able to make use of that richly endowed armoury of weapons useful in the stereoselective elaboration of cyclic systems, to which I alluded in my opening remarks. Then, *by desulphurisation, V → VIII (see Fig. 5), we should create a characteristic array of alternately spaced methyl groups, attached to a now flexible linear chain* (cf. Fig. 5 — VIII).

DECISION Nr. 2:

    OBLITERATE THE INTRACTABLE STEREOCHEMICAL PROBLEMS ASSOCIATED WITH GENERATION AND AGGRANDIZEMENT OF CHIRALITY IN NON-RIGID SYSTEMS

By

    CREATIVE INTRODUCTION OF ELEMENTS OF RIGIDITY WHICH PERMIT STEREOSPECIFIC OPERATIONS OF THESE TYPES:

        a. KNOWN

        b. PREDICTABLE

        c. DESIRED, WHETHER OR NOT KNOWN OR PREDICTABLE

Fig. 4.

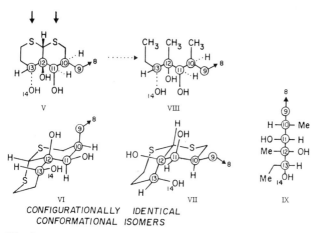

Fig. 5.

In the specific case illustrated, the result would be an intermediate representing the C.9 through C.13 chain of erythromycin, appositely substituted and correctly oriented, both in absolute and relative senses, as we may discern through inspection and correlation of VIII with our monitor IX (Fig. 5). Thus, the 'creative introduction of elements of rigidity' boldly adumbrated in Fig. 4 finds a specific expression in our plan in the use of sulphur atoms to create rings by bridging carbon atoms destined to become methyl groups; indeed, a glamourisation of the lowly, usually modest methyl group, which at first sight would hardly be expected to play a prominent role in the direction of stereoselective synthetic operations! In relation to the importance of the role of imagination in synthetic planning, it is worthy of note that there is nothing in the erythromycin structure *per se* which, either explicitly or implicitly, suggests the basic plan of attack we adopted.

All right; our planes have been laid! We may now proceed to the laboratory bench. Our saga begins with the simple, readily available tetrahydro-γ-thiapyrone X (Fig. 6), which was readily converted by methyl *ortho*formate and methanol in the presence of toluene sulphonic acid to the ketal XI (Fig. 6), and thence, by a very smooth sequence of substitution and hydrolysis reactions, to the cyclic dithiohemiacetal XII (Fig. 6).

In order to prepare the intermediate needed for reaction with XII, the path shown in Fig. 7 was laid down. Methyl-γ-benzyloxybutyrate (Fig. 7 – XIII), when treated with excess methyl formate in the presence of lithium di-isopropylamide at $-78°$, gave the formyl derivative [XIV (Fig. 7)], which afforded the

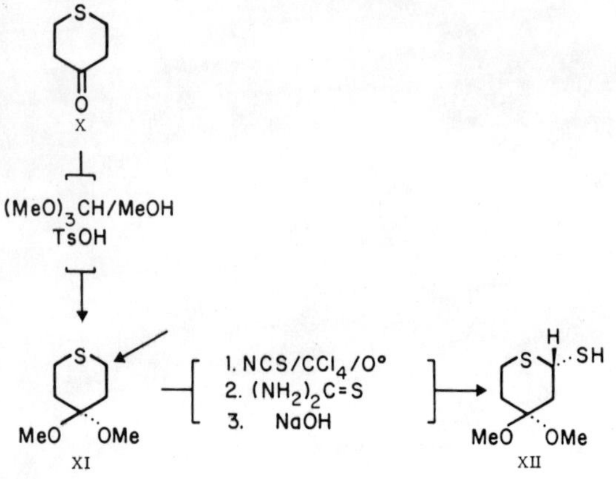

Fig. 6.

acetal XV (Fig. 7) upon treatment with methyl *ortho*formate and methanol in the presence of sulphuric acid. Lithium aluminium hydride in ether brought about the reduction of the carbomethoxy group in XV, and afforded the carbinol XVI (Fig. 7), which in its turn was mesylated in the usual way to give the desired intermediate XVII (Fig. 7).

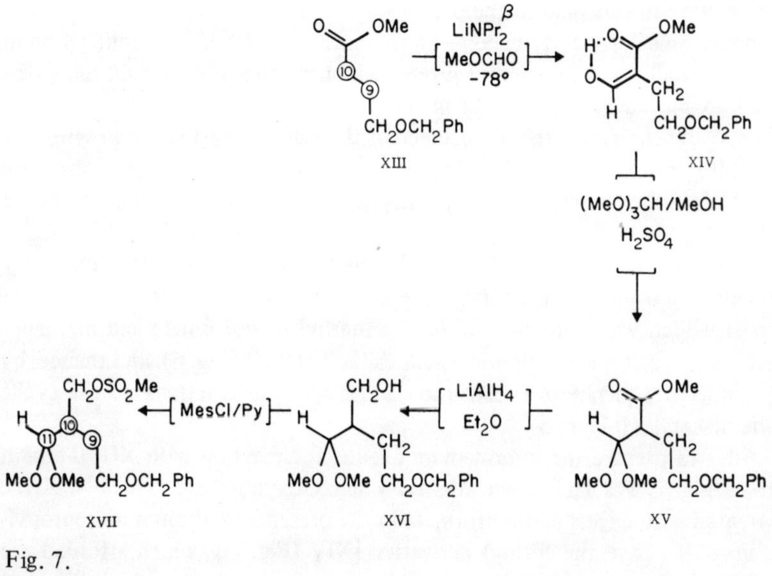

Fig. 7.

With the mercaptan XVIII (Fig. 8) = XII (Fig. 6) and the mesylate XIX (Fig. 8) = XVII (Fig. 7) in hand, it was gratifying to find that they underwent smooth condensation, in methanol solution in the presence of sodium methoxide, to give XX and XXI (Fig. 8). Since, at this stage, both XVIII and XIX contain an asymmetric centre, and are racemic, the condensation products XX and XXI are of necessity racemic diastereomers, of which, on behalf of simplicity, only one enantiomer is shown; we shall continue this simplified representation, until, at the appropriate point, we take up the discussion of optically active intermediates.

In practice, the diastereomers XX (Fig. 8) = XXII (Fig. 9) and XXI = XXIII (Fig. 9) were not separated; rather, the mixture was subjected to complete deketalisation and deacetalisation, using aqueous acetic acid, to give the ketoaldehydes XXIV and XXV (Fig. 9) With silica gel as catalyst, these substances were found to undergo very clean aldolisation, yielding XXVI and XXVII (see Fig. 9), *as sole products*, which were readily separable by simple chromatographic methods, and obtained in the pure crystalline state, in approximately equal amounts. In each of these aldolisation reactions, two new asymmetric centres are created, and our observations demonstrate that these centres are created *stereospecifically*. Can we discern the bases for these most welcome results?

The picture XXVIII (Fig. 10) presents a *quasi* three-dimensional sketch of the approximate geometrical relationships which must obtain in the transition state for the aldolisation reactions. Clearly the ketonic carbonyl group must be enolised; the resulting double-bond-containing six-membered rings will adopt a quasi-chair conformation. Now – and of most importance – the *anomeric effect*, which is well known for oxygen compounds, and clearly is operative here in these sulphur analogues, causes the side-chains attached to the rings by

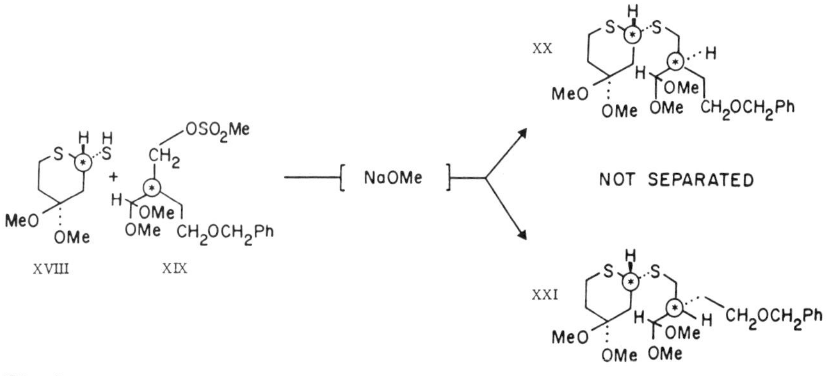

Fig. 8.

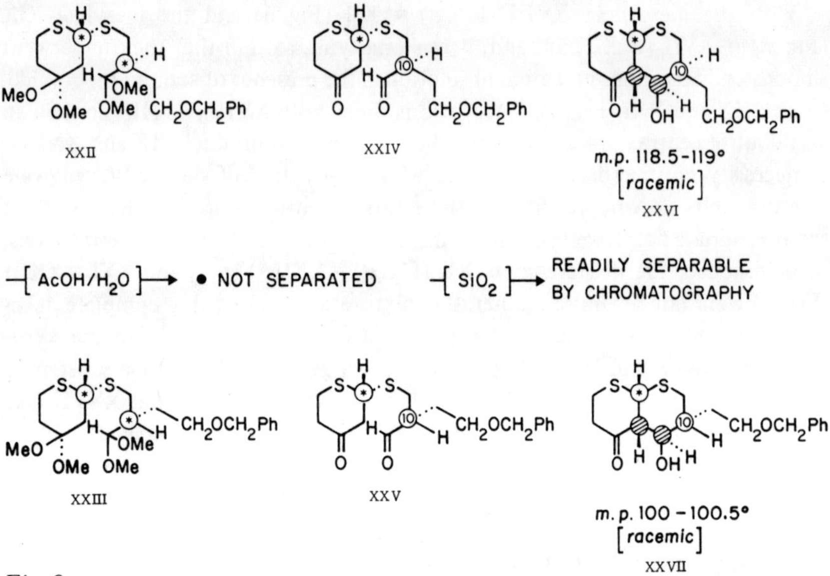
XXII
XXIV
m.p. 118.5-119°
[racemic]
XXVI
AcOH/H2O
• NOT SEPARATED
SiO2
READILY SEPARABLE BY CHROMATOGRAPHY
XXIII
XXV
m.p. 100-100.5°
[racemic]
XXVII

Fig. 9.

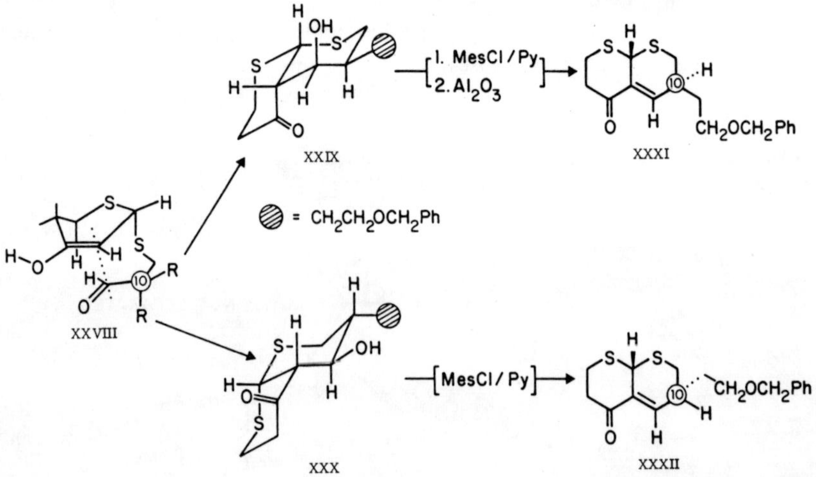
XXIX
1. MesCl/Py
2. Al2O3
XXXI
= CH2CH2OCH2Ph
XXVIII
XXX
MesCl/Py
XXXII

Fig. 10.

sulphur atoms to adopt quasi-axial orientations. In these circumstances, cyclisation can *only* lead stereospecifically to *cis*-fused dithiadecalin systems; indeed, both our sulphur atoms are serving us well! The observed β orientation of the newly-formed hydroxyl groups in both aldols is explicable in more pedestrian terms; the aldehyde oxygen atoms simply orient themselves in the least sterically crowded positions.

Each of the aldols, represented three-dimensionally in XXIX and XXX (Fig. 10), adopts a different one of the two conformations a priori possible for *cis*-dithiadecalin derivatives; the choice is of course determined by the large $-CH_2CH_2OCH_2Ph$ groups, which in each case adopts an equatorial orientation.

Either aldol XXIX or XXX, was found to be convertible to the corresponding α,β-unsaturated ketone (Fig. 10 – XXXI or XXXII), by mesylation, followed by elimination. The reactions were rather more sluggish, though no less clean, in the case of XXIX, for which we adopted the household sobriquet aldol I, the more readily to distinguish it from its congener XXX, designated aldol II. It is important now to make two points of special interest. First, the structural, stereochemical and conformational details presented here were initially assigned on the basis of numerous systematic physical measurements, which need not be presented here, though it may be mentioned that proton magnetic resonance studies played the most definitive role. Second, careful scrutiny of stereochemical relationships reveals that XXXI (Fig. 10) is the diastereomer whose further elaboration may be expected to lead to substances having the stereochemical relationships present in erythromycin at the appropriately related centres; consequently the sequences leading to, and from XXXI may be designated the 'natural' series. In the sequel, we shall be concerned in the main with substances of this natural series.

The reduction of the ketone XXXIII (Fig. 11) = XXXI (Fig. 10), using sodium borohydride in methanol/dimethoxyethane at $0°$, led, as expected, stereospecifically, to the α-oriented alcohol XXXIV (Fig. 11), which was acetylated without incident to XXXV (Fig. 11). Consideration of models, as well as ample precedent in carbocyclic systems, gave us every reason to suppose that osmylation of XXXV would involve attack on the top face of the molecule, and such indeed was the case. Osmium tetroxide in ether solution, followed by sodium bisulphite, gave the acetoxy diol (Fig. 1 – XXXVI) as sole product. Now, when XXXVI was treated with acetone dimethyl ketal in the presence of toluene sulphonic acid, the acetoxy ketal XXXVII was produced in high yield.

With the obtention of XXXVII (Fig. 12) = XXXVII (Fig. 11), we supposed that we had in hand an intermediate in which all of the problems of relative stereochemistry at C.10, C.11, C.12 and C.13 of the erythromycin main chain had been solved, as reference to our monitor XL (Fig. 12) will confirm. But, we

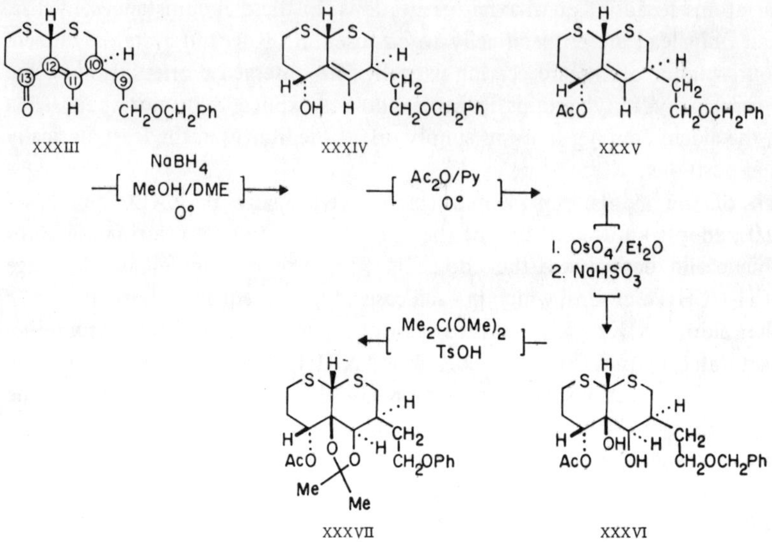

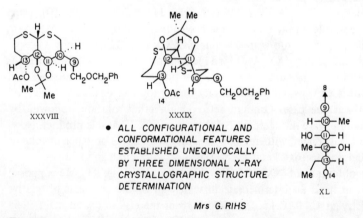

Fig. 11.

were now some distance from home, and still faced a long journey, which would best be launched from an absolutely firm base. Were our stereochemical arguments and deductions correct? Confident though we might have been, we were certainly delighted to have unequivocal proof in the shape of a three-dimensional X-ray crystallographic analysis carried out for us in Basel by Mrs. G. Rihs. Her results, shown in Fig. 12 (XXXIX) confirmed our expectations in every detail.

Fig. 12.

So far, all of the work I have described was carried out with racemic substances. It is now time to turn to the extension of our studies to optically active materials. In Fig. 13, we portray both enantiomers XLI and XLII (Fig. 13), which together comprise the now familiar racemic monocyclic dithiohemiacetal; of these, XLI possesses the *absolute* configuration necessary for elaboration into erythromycin of the correct, natural absolute configuration. The racemic substance was converted by reaction with (−)-camphanylchloride XLIII (Fig. 13) to a mixture of the diastereomeric esters XLIV and XLV (Fig. 13), from which XLIV was readily separated, optically pure, by crystallization. Again, through the kindness of Mrs. Rihs, the structure of this substance was rigorously established in every detail through three-dimensional X-ray crystallographic analysis. In this case, since the absolute configuration of (−)-camphanic acid is known, the structure determination *established unequivocally the absolute configuration of the cyclic dithiohemiacetal moiety.* It remained only to effect the cleavage of XLIV, by sodium methoxide catalysed methanolysis, to obtain XLI, in the optically pure state, and of the desired absolute configuration.

The apparent simplicity of the effective sequence just described in fact conceals a highly useful discovery which could hardly have been anticipated: *cyclic dithiohemiacetals of the type XLI (Fig. 13) are configurationally stable.* Such is

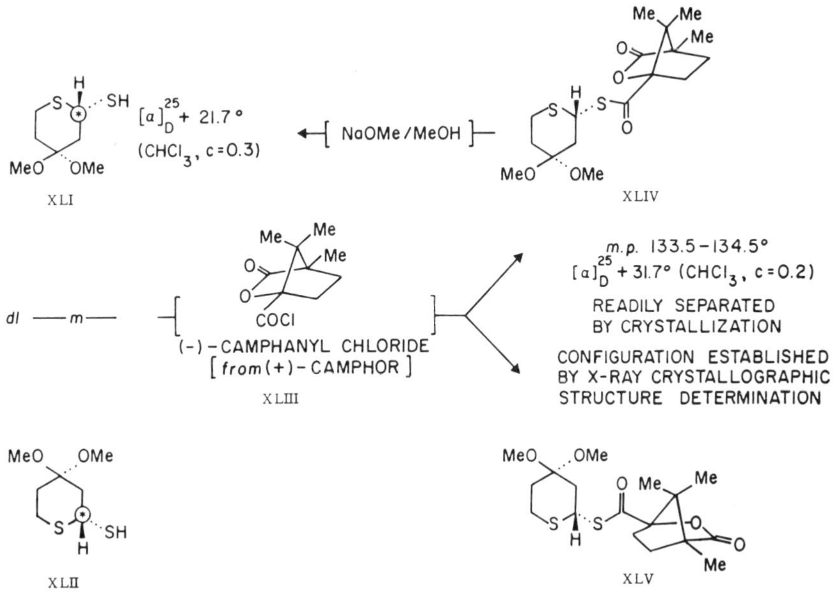

Fig. 13.

most certainly not the case with oxygen analogues, which are notoriously susceptible to loss of stereochemical integrity at the anomeric, asymmetric carbon atom, through participation in ring-chain equilibria, involving non-chiral tautomers. When we recognise that it is the strikingly different behaviour of the sulphur compounds which enabled us to initiate our studies in the optically active series in a most simple and efficient manner, we have excellent reason to be pleased with the fact that the 'creative introduction' of sulphur atoms in our planning yielded benefits well beyond those we had hoped for.

Now we were in a position to follow the path clearly laid down in our studies in the racemic series. The optically active dithiohemiacetal XLVI (Fig. 14) = XLI (Fig. 13) was condensed, as before, with the *racemic* mesylate XLVII + XLVIII (Fig. 14), to give a mixture, now of *optically active* diastereomers XLIX and L (Fig. 14). Again, this mixture, of LI (Fig. 15) = XLIX (Fig. 14) and LII (Fig. 15) = L (Fig. 14), was not separated at this stage, but rather, subjected to mild hydrolysis by aqueous acetic acid to the ketoaldehydes LIII and LIV (Fig. 15). The latter, in their turn, were subjected to aldolisation, *now catalysed by D-proline,* to produce, completely stereospecifically, aldol I − LV and aldol II − LVI (Fig. 15) in the optically active state. As before, the aldols were readily separable by chromatography, and obtained optically pure by simple crystallization. Thus, the pure aldol LV (Fig. 15) of the desired 'natural' *absolute* configuration was now available for subsequent operations.

Following the procedure established in the racemic series, the optically active

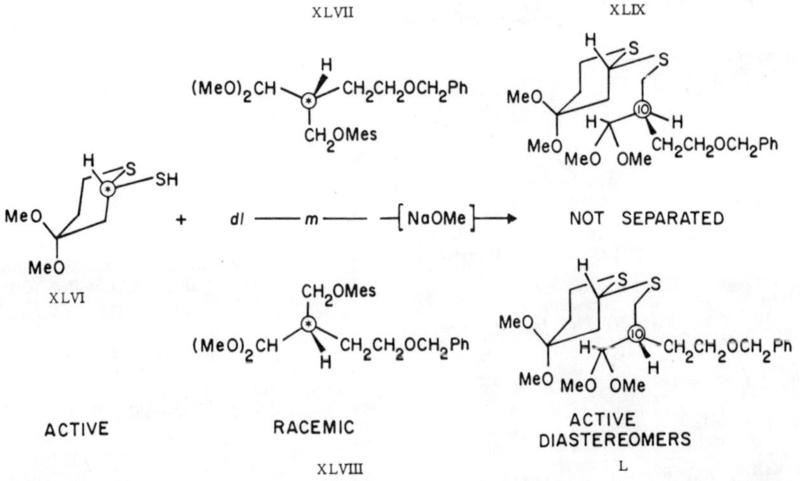

Fig. 14.

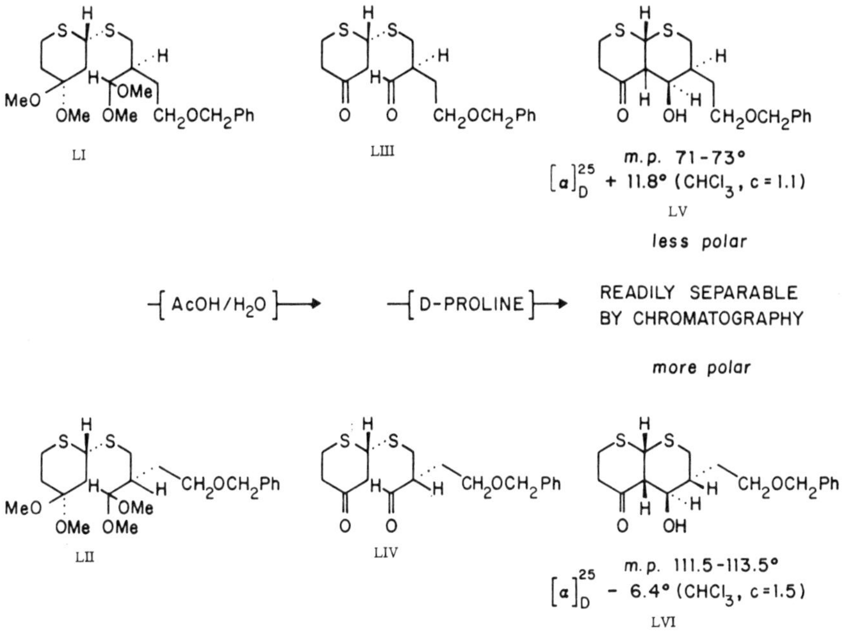

Fig. 15.

[Structures LI, LIII, LV (m.p. 71-73°, $[\alpha]_D^{25}$ + 11.8° (CHCl$_3$, c = 1.1), less polar)]

―{ AcOH/H$_2$O }→   ―{ D-PROLINE }→   READILY SEPARABLE BY CHROMATOGRAPHY

more polar

[Structures LII, LIV, LVI (m.p. 111.5-113.5°, $[\alpha]_D^{25}$ − 6.4° (CHCl$_3$, c = 1.5))]

aldol I, LVII (Fig. 16) = LV (Fig. 15), was converted to the α,β-unsaturated ketone LVIII (Fig. 16), using mesyl chloride and pyridine, followed by alumina. The ketone LVIII was next osmylated directly in ether solution, through the agency of osmium tetroxide/pyridine, and the resulting osmium complex was decomposed by sodium bisulphite, to give the diol LIX (Fig. 16) as sole product. Treatment of the latter with acetone dimethylacetal in the presence of toluene sulphonic acid proceeded without incident, to yield the ketal LX (Fig. 16); we shall see shortly that these reactions complete the formation of one of our key intermediates.

The ketone LXI (Fig. 17) = LX (Fig. 16) now served as the starting point for a new series of transformations. First, it was reduced, by sodium borohydride in methanol/dimethoxyethane at 0°, to the alcohol LXII (Fig. 17), which, using potassium hydride, tetrahydrofuran and methyliodomethyl ether, was converted to the methoxymethyl ether LXIII (Fig. 17).

We approached the next operation with some trepidation, for a crucial element in our plan was about to be put to the acid test. The sulphur atoms, which

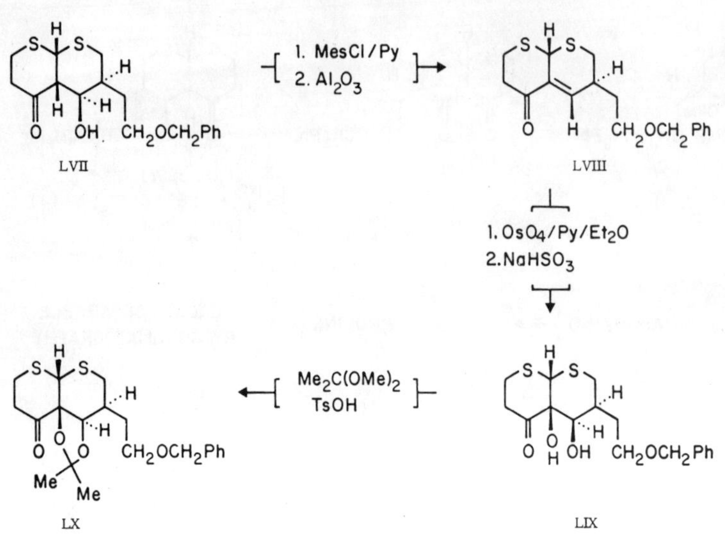
1. MesCl/Py
2. Al_2O_3
1. OsO_4/Py/Et_2O
2. $NaHSO_3$
$Me_2C(OMe)_2$
TsOH
LVII
LVIII
LIX
LX

Fig. 16.

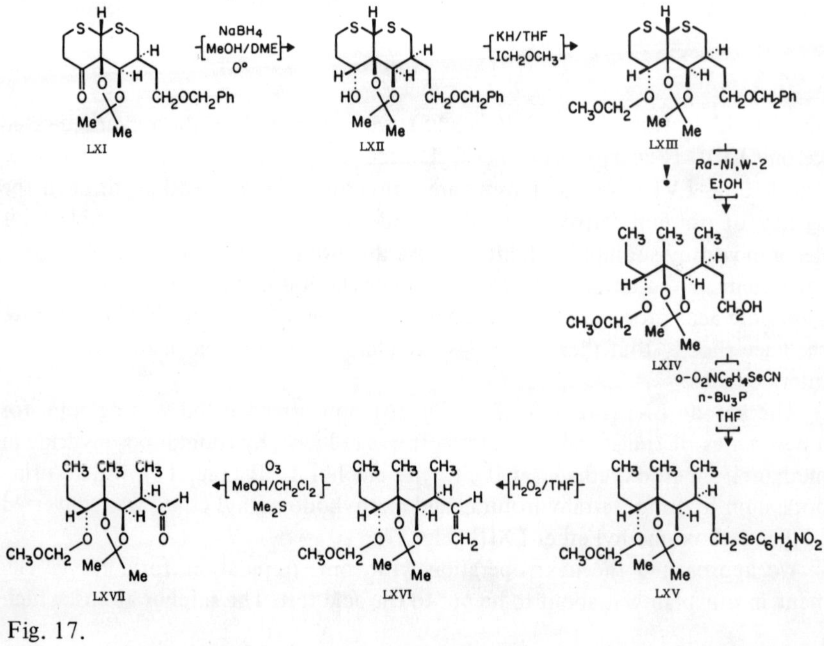
$NaBH_4$
MeOH/DME
0°
KH/THF
ICH_2OCH_3
Ra-Ni,W-2
EtOH
o-$O_2NC_6H_4SeCN$
n-Bu_3P
THF
H_2O_2/THF
O_3
MeOH/CH_2Cl_2
Me_2S
LXI
LXII
LXIII
LXIV
LXV
LXVI
LXVII

Fig. 17.

had played their several roles so admirably, had but one more task to perform: they must retire gracefully from the scene, leaving in the wake of their departure an array of methyl groups required in our objective. And so, in the event, they did! When the dithiadecalin LXIII was treated with W-2 Raney nickel in ethanol, exceptionally smooth and uncomplicated desulphurisation took place, with the formation of LXIV (Fig. 17). The desired newly installed methyl groups manifested themselves most clearly in the beautiful proton magnetic resonance spectrum of LXIV — a textbook example, in which three resonances, appropriately split, for methyl groups bound, respectively, to secondary, tertiary and quaternary carbon atoms were clearly to be seen. Concomitantly with the desulphurisation, and most felicitously, the benzyl group of LXIII was cleaved, leaving the desired free primary carbinol grouping of LXIV.

The selenium compound LXV (Fig. 17) was cleanly produced when the alcohol LXIV was treated with o-nitrophenylselenocyanate and tri-n-butylphosphine in tetrahydrofuran. Oxidation of LXV with hydrogen peroxide in tetrahydrofuran then led directly to the olefin LXVI (Fig. 17), no doubt through the intermediacy of a fugitive selenoxide. In its turn, the olefin was smoothly cleaved by ozone in methanol/methylene chloride, followed by dimethylsulphide, to give the aldehyde LXVII (Fig. 17) — the second of our desired key intermediates.

The time is now at hand to direct attention to a highly significant singularity of the erythromycin structure. Inspection of LXX (Fig. 18) reveals that the stereochemical relationships and substitution patterns, at C.4, C.5 and C.6 on the one hand, and at C.10, C.11 and C.12 on the other, are identical! It is this remarkable fact that has enabled us to synthesize *both* key intermediates — the ketone LXVIII (Fig. 18), representing C.4, C.5 and C.6, and the aldehyde LXIX (Fig. 18), representing C.10, C.11, C.12 and C.13, *by a common path*. Note further that the complementarity of functionality within the key intermediates — at C.8 in the ketone, and at C.9 in the aldehyde — is such that the stage is set for their union, with formation of the required C.8–C.9 bond.

And indeed, when the anion derived from the ketone LXVIII by treatment with mesityl lithium in tetrahydrofuran at $-78°$ was allowed to react with the aldehyde LXIX at $-50°$, smooth condensation occurred, yielding a mixture of diastereomeric aldols LXXI (Fig. 19). The slight stereochemical complication at this stage was of no consequence, since it was expunged forthwith, when the aldol mixture was oxidised — using trifluoroacetic anhydride/dimethyl sulphoxide in methylene chloride, in the presence of diisopropylethylamine — to the diketone LXXII (Fig. 19), which, it is of some interest to note, exists to some extent in the un-enolised form.

It was now necessary to remove the unwanted oxygen atom at C.7 [cf. LXXII

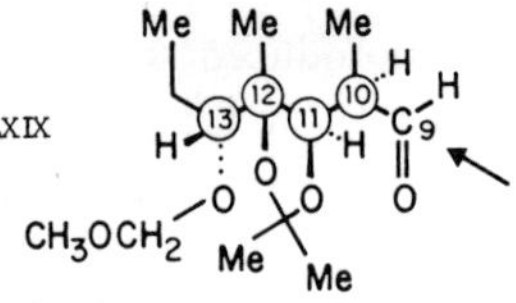

Fig. 18.

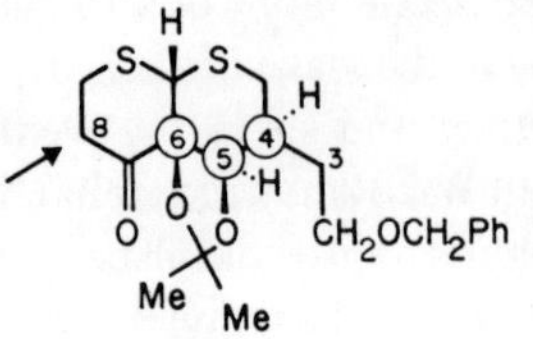

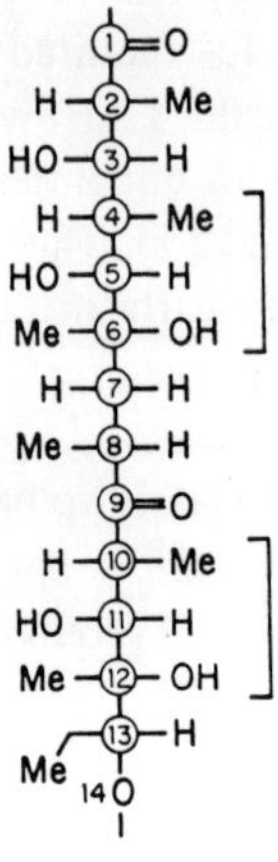

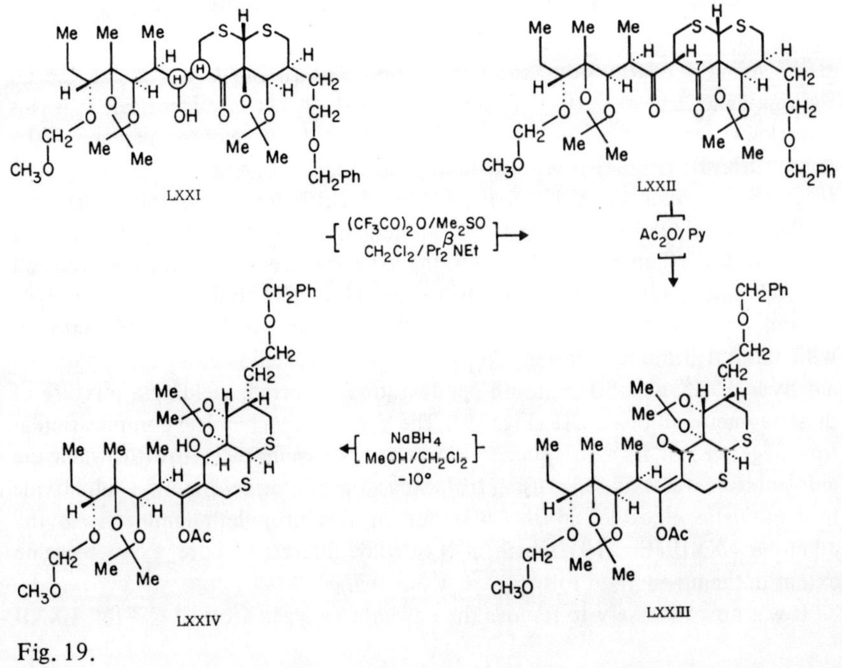

Fig. 19.

and LXXIII (Fig. 19)]. To lay the basis for this change, a study in some depth was made, of the enolisation, acylation and reduction of β-dicarbonyl compounds. Surprisingly, in view of the fact that this arena has been a playground of organic chemists for a hundred years, new and highly interesting observations were made, which could themselves form the basis of a separate lecture. For our present purpose, suffice it to say that LXXII was converted by acetic anhydride in pyridine, under carefully defined conditions, to the *trans* enol acetate LXXIII, which in its turn, was reduced by sodium borohydride in methanol/methylene chloride at $-10°$ to the alcohol LXXIV (Fig. 19).

And now, the hydroxy enolacetate LXXV (Fig. 20) = LXXIV (Fig. 19) was converted by mesyl chloride in pyridine to the α,β-unsaturated ketone LXXVI (Fig. 20). The unwanted oxygen atom had been removed, and the conjugated carbon-carbon double bond of LXXVI was susceptible of ready reduction. In the formation of the saturated product LXXVII (Fig. 20), asymmetry is regenerated at C.8, *in the desired sense*. This outcome is a consequence of a rather pretty feature of our plan: LXXVI is a *cis* fused dithiadecalin, and attack by reagents may be expected to take place preferentially on the convex face of the molecule.

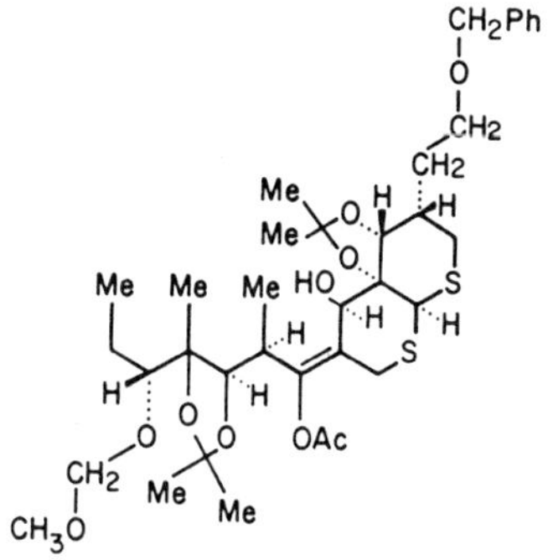

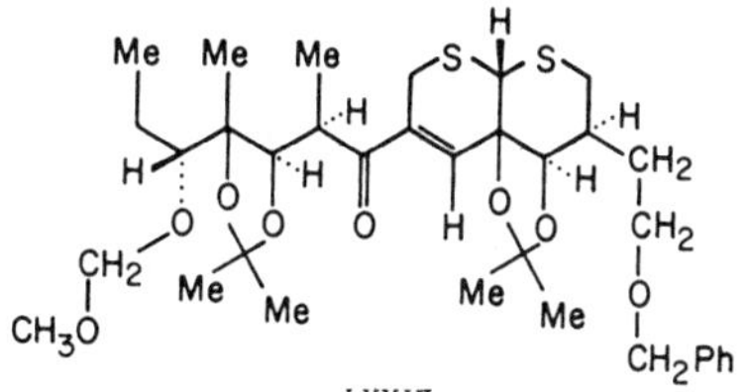

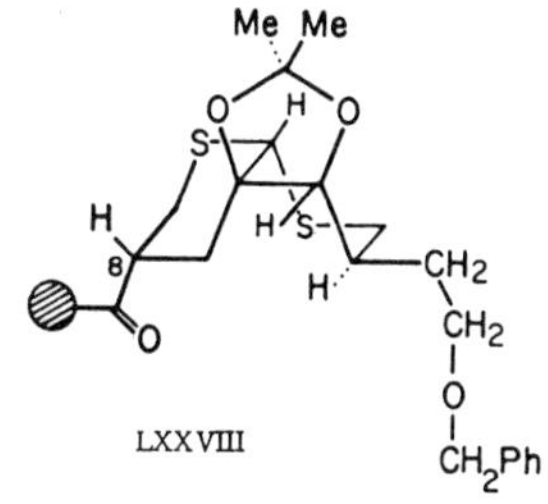

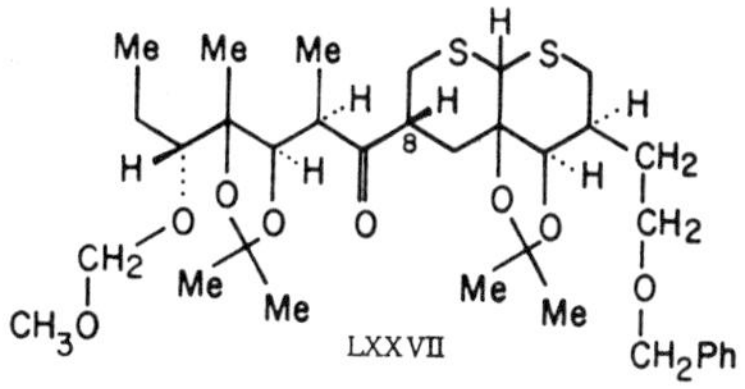

Fig. 20.

Beyond that, the product LXXVII will adopt that one of the two a priori available conformations in which the massive $-CH_2CH_2OCH_2Ph$ substituent is disposed in the equatorial sense, as shown in LXXVIII (Fig. 20); it is now evident that the stable orientation at C.8 must be the desired one, in which the large attached group is also equatorial.

With the obtention of LXXIX (Fig. 21) = LXXVII, eight of our asymmetric centres are in place, appositely substituted, and properly oriented, in respect to one another, as well as in the absolute sense. How shall we now proceed? Clearly, the side chain $-\overset{3}{C}H_2CH_2OCH_2Ph$ must be modified. We feel confident that the methods which served us so well (see Fig. 17) in the synthesis of the aldehyde LXXX (Fig. 21) will enable us to convert LXXIX into an analogous substance, in which C.3 enjoys the status of the carbon atom of an aldehyde group. Further, it is now apparent that the aldehyde LXXX, which was used hitherto for the construction of the C.9 → C.13 segment of LXXXII (Fig. 21), is also an excellent model for studies of the elaboration of the still needed C.1 → C.3 moiety. And indeed, when the anion prepared from ethyl propionate by the action of lithium diisopropylamide in tetrahydrofuran is treated with LXXX at $-78°$, the hydroxyester LXXXI (Fig. 21) is produced. The stereochemistry of this process is very highly selective in respect of the creation of asymmetry at C.3, but approximately an equal amount of a diastereomer, epimeric at C.2, is

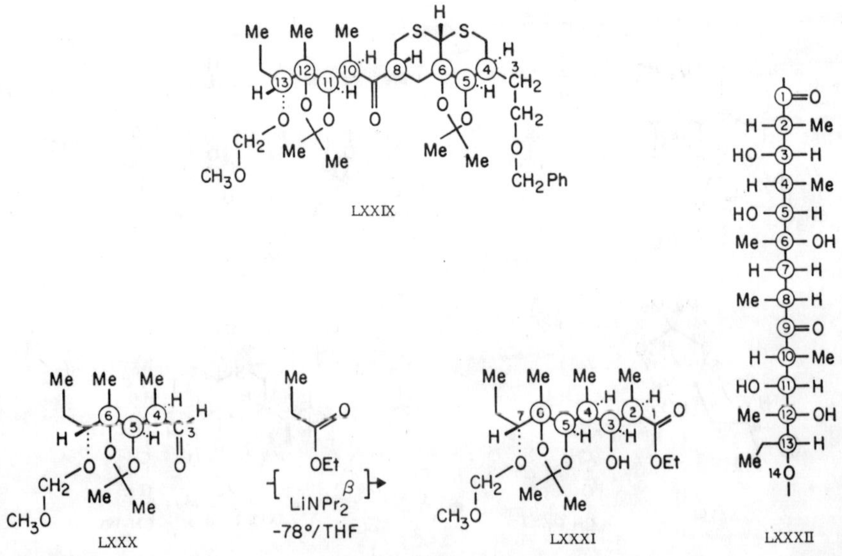

Fig. 21.

formed. Certainly this route may be regarded as flawed in the stereochemical sense. But, its extreme simplicity gives it an immediate practical advantage which might be surpassed only with very great difficulty by more sophisticated multi-stage processes.

Taken altogether, these considerations and observations lead us to suppose that our first major objective – the construction of the entire chain LXXXII is well within sight. Of course, we are aware that we may encounter ancillary problems, such as the possible necessity to mask temporarily the carbonyl group at C.9, and the maintenance of stereochemical integrity at C.8 and C.10. But none of these looms in our minds as an especially difficult obstacle.

If our presentiments are not over-optimistic, we shall in the not distant future be able to turn our major attention to the closure of the fourteen-membered lactone ring, and the attachment of the sugar residues, with the object of completing a total synthesis of erythromycin.

Work of the kind I have described is highly demanding of those who do it. If you have enjoyed with me this portrayal of our chemical adventure, your indebtedness is, as is mine, to the men and women whose names appear in Fig. 22. Their skill, devotion and application are beyond praise.

1972/73	1973/74	1974/75	1975/76	1976/77
C. H. CHEN				
P. BALARAM				
Y. KOBUKE				
W. HEGGIE				
S. MALCHENKO				
	A.T. VASELLA			
	P.A. JACOBI			
	J.A. HYATT			
	K. TATSUTA			
	J.B. PRESS			
	E.A. TRUESDALE			
		H.-J. GAIS		
		H.M. SAUTER		
		K. KOJIMA		
		P.A. WADE		
			L.M. TOLBERT	
1977/78			Y. UEDA	
			D. IKEDA	
P. J. CARD			V.J. LEE	
T. V. RAJAN BABU			M. SUZUKI	
T. LEUTERT		1978 –	T. UYEHARA	
K. SAKAN			R. B. CHÊNEVERT	
B. S. ONG		R.S. MATTHEWS	D. G. GARRATT	
D.W.H. HOPPE		L.J. BROWNE		
I. HOPPE		W. C. VLADUCHICK		D. P. HESSON
K. HAYAKAWA				J. MARTENS
B.-W. AU-YEUNG				H. N.– C. WONG
				I. UCHIDA

Fig. 22.

Acknowledgement

All of us are particularly indebted to the National Institute of General Medical Sciences for generous financial support [GM04229].

CHAPTER 4

Synthesis and antiviral activities of new 5-substituted pyrimidine nucleoside analogs

PAUL F. TORRENCE, ERIK DE CLERCQ, JOHAN DESCAMPS, GUANG-FU HUANG and BERNHARD WITKOP

One of the major challenges of the interdisciplinary field at the border of organic chemistry and virology is the rational development of selective antiviral agents. Nucleoside analogs that are structurally related to the intermediates of RNA and DNA biosynthesis can be expected to interfere with viral-biogenesis, especially so since virus replication places increased demands on the host cell for the synthesis of nucleic acids. However, experience from the use of nucleoside analogs in the therapy of neoplastic diseases suggests that complete reliance on the increased burden of DNA and/or RNA synthesis in the malignant or virus-infected cell will not provide a sufficient measure of selectivity to give a chemotherapeutically selective agent. Any cell that is actively involved in nucleic acid manufacture, whether the cell be normal, malignant or virus-infected, will be more or less susceptible to the effects of such nucleic acid antagonists. Thus, many such agents have clearly undesirable effects on the host, such as immunosuppression and embryotoxicity. Furthermore, the great peril associated with the use of nucleoside analogs is that they may be mutagenic or carcinogenic due to incorporation into the genes of the host cell.

Fortunately, there is now considerable evidence [1–8,77] that for a number of viruses, certain critical enzymes involved in nucleic acid synthesis are (a) newly induced by the infecting virus's genome and (b) these enzymes often differ significantly in their properties compared to those of the host cell counterparts. Exploitation of these differences in the level and characteristics of the virus-induced enzymes could, therefore, provide the required degree of selectivity to yield a clinically useful nucleoside antiviral agent.

With this information at hand, two different approaches might be envisioned. The first is to isolate and characterize thoroughly all virus-induced enzymes and determine if they can be selectively inhibited in comparison to the corresponding host cell counterpart. The second approach is to explore rationally defined new nucleoside analogs to learn whether or not subtle structural changes can effect differences in relative selectivity of virus-infected compared to uninfected cells. Although both approaches are being followed in different laboratories, we will be concerned here with only the second, more chemical, avenue of exploration.

The best known nucleoside antiviral agent is 5-iodo-2′-deoxyuridine (IUdR). This analog has pioneered much of our current understanding of virus replication and has pointed to the promise of nucleosides as antiviral agents; nonetheless, it possesses the important limitations of immunosuppression (in vitro), embryotoxicity, mutagenicity (in phage and eukaryotic cells), and activation of oncornavirus particles in mammalian cell culture. The mechanism of antiviral action and the mechanism of its toxic effects (mutagenicity, oncornavirus activation) may be due to incorporation into viral and host cell DNA, respectively. Once IUdR is incorporated in place of thymidine into DNA, the effect of the replacement of methyl by iodine could have relatively drastic effects on DNA including (a) alteration of the keto-enol tautomerism with a resultant greater frequency of base-mispairing during DNA replication or repair and (b) changes in DNA-repressor protein interactions.

On the basis of such considerations, some nucleoside analogs have been synthesized with a view to suppressing either (a) the mutagenicity of the DNA-incorporated 'fraudulent' nucleoside or (b) blocking incorporation into DNA altogether. The first approach is exemplified by 5-ethyl-2′-deoxyuridine (EtUdR) [9,10]. The pK_a value for the dissociation of N–3–H of EtUdR is the same as the corresponding value for thymidine. It's introduction into DNA would not be expected to change the keto-enol tautomerism of the base as compared to thymidine, and thus no increased mutation frequency should occur. Indeed, while EtUdR possesses significant in vitro and in vivo antiviral activity [78,79] it is non-mutagenic in phage and does not activate oncornavirus production in mammalian cell culture [10,76]. The second approach, that of modifying the 5′–OH group to prevent incorporation into DNA, has been explored by Prusoff and his coworkers. Of a number of different 5′-modified nucleoside analogs 5′-amino-5-iodo-2′,5′-dideoxyuridine was shown to possess specific antiherpes virus activity [11,12,80]. This analog, AIdU, is non-toxic to uninfected cells. In the infected cell herpes simplex virus induces high levels of thymidine kinase leading to the phosphorylation of AIdU and its resultant selective toxicity for virus-infected cells only. AIdU is incorporated into DNA of herpes-virus-infected cells, implying the formation of a phosphoramidate bond in the nucleic acid [13]. The 5′-N-triphosphate of AIdU converts E. coli thymidine kinase into an inactive dimer. In fact, this N-triphosphate of AIdU is a more potent allosteric inhibitor of the kinase than either thymidine triphosphate or 5-iodo-2'-deoxyuridine 5'-triphosphate [14].

Variation of the physical properties of 5-substituted 2'-deoxyuridines

We wished to ask the question: What effect do different 5-substituents have on the antiviral activity of 2'-deoxyuridine? To this end, we prepared a number of analogs with a view to maximizing the electronic, steric and lipophilic differences among the introduced substituents. Table 1 shows how two such properties that reflect these differences vary among the newly synthesized analogs and some other nucleosides of relevance. The lipophilic character of the new analogs range from the carboxamidomethoxy (least lipophilic) to the benzyloxy substituent (most lipophilic) with the (natural) methyl group (thymidine) occupy-

Table 1. Effect of the nature of the 5-substituents on the pyrimidine N-3-H pK_a and nucleoside partition coefficient [15].

R	N-3-H pK_a^a	log P_{oct}^b
H–	9.3 [c]	–1.526
CH_3–	9.8	–1.140
I–	8.2 [d]	–0.609
F–	7.8 [c]	–1.30
CN–	6.8	–1.57
NO_2–	5.8	–
HC≡C–CH_2–O–	8.4	–1.09
O ⫽ C–CH_2O– | NH_2	8.3	–2.49
NCS–	–	–1.06
$C_6H_5CH_2$–O–	8.5	0.112

[a] Apparent pK_a's were determined spectrophotometrically using Teorell-Stenhagen buffers.
[b] As defined by the equilibrium nucleoside (H_2O) ⇌ nucleoside (octanol) with

$$P = \frac{\text{concentration in organic phase}}{\text{concentration in aqueous phase}}$$. Determined at 25° in phosphate buffer

(pH 7) according to techniques outlined in reference [16]. The more negative the log P_{oct}, the more hydrophilic is the nucleoside.
[c] As determined by Berens and Shugar [17].
[d] This study [15] and [17].

ing an intermediate position. The electronic effect of the introduced substituent is reflected in the N–3–H pK_a's which range from 5.8 for 5-nitro-2′-deoxyuridine (>95% ionized at pH 7.2) to 8.4 and 8.5 for 5-propynyloxy- and 5-benzyloxy-2′-deoxyuridine, respectively (both ~5% ionized at pH 7.2). The range of properties surveyed in Table 1 may be reflected in the biological properties of these nucleosides.

5-Thiocyano pyrimidine nucleosides [18,19]

The synthetic approach to 5-thiocyano pyrimidine nucleosides [*18,19*] was based on the thiocyanation of olefins [*20,21*], aromatic compounds [*22–24*] and pyrole carboxylates [*25*] by chlorothiocyanogen (CISCN). We found that uridine, 2′-deoxyuridine, uracil arabinoside, tri-O-chloroacetyl uridine, triacetyl uridine as well as uracil itself reacted smoothly with chlorothiocyanogen in glacial acetic acid to give the corresponding 5-thiocyano derivative in moderate to excellent yields. Furthermore, these nucleosides provided a convenient access to 5-mercapto pyrimidine nucleosides, since they could be easily reduced, e.g. with dithiothreitol or glutathione (Scheme 1).

Determination of the biological activity of these 5-thiocyanopyrimidine nucleosides was of special interest, since such analogs might behave in one of two different manners: (a) as pseudohalogen derivatives, with the SCN group mimicking the iodo group present in the clinically useful antiviral agent IUdR; (b) the thiocyano derivatives might be reduced in vivo to the 5-mercapto analogs. Such a reduction seemed quite plausible in view of the facile reduction in vitro of such analogs by glutathione [*19*], an ubiquitous constituent of living systems [*26*]. This second pathway would lead to products with established biological activity, since both 5-mercaptouridine and 5-mercapto-2′-deoxyuridine [*27–29*] had been previously synthesized and demonstrated to possess both antibacterial and antitumor properties [*30,31*].

5-Cyanopyrimidine nucleosides

Two basically different chemical approaches have been used to prepare 5-cyano-substituted pyrimidine nucleosides. The first relies upon the condensation of a base or a base precursor and a suitably protected ribose derivative. Shaw et al. [*32*] condensed tri-O-benzoylated D-ribosylamine with α-cyano-β-ethoxy-N-ethoxycarbonylacrylamide to yield, after deblocking, 5-cyanouridine which could also be synthesized by a similar condensation using an isopropylidene derivative of furanosylamine. The Hg(CN)$_2$-nitromethane procedure was used by Watanabe and Fox [*33*] to obtain a 70% yield of the tribenzoate of 5-cyano-

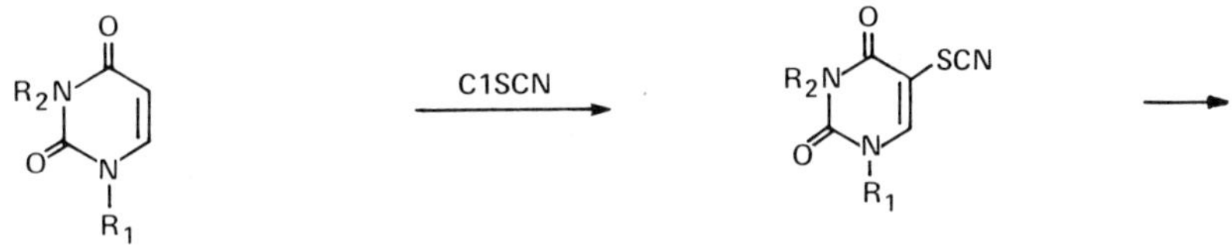

3, R_1 = CH_3; R_2 = CH_3
5, R_1 = 2', 3', 5'-tri-O-chloroacetyl-β-D-ribofuranosyl; R_2 = H
7, R_1 = β-D-ribofuranosyl; R_2 = H
8, R_1 = 2'-deoxy-β-D-ribofuranosyl; R_2 = H
9, R_1 = β-D-arabinofuranosyl; R_2 = H
12, R = 2', 3', 5'-tri-O-acetyl-β-D-ribofuranosyl; R_2 = H

4, R_1 = CH_3; R_2 = CH_3
6, R_1 = β-D-ribofuranosyl; R_2 = H
10, R_1 = 2'deoxy-β-D-ribofuranosyl; R_2 = H
11, R_1 = β-D-arabinofuranosyl; R_2 = H
13, R_1 = 2', 3' 5'-tri-O-acetyl-β-D-ribofuranosyl; R_2 = H

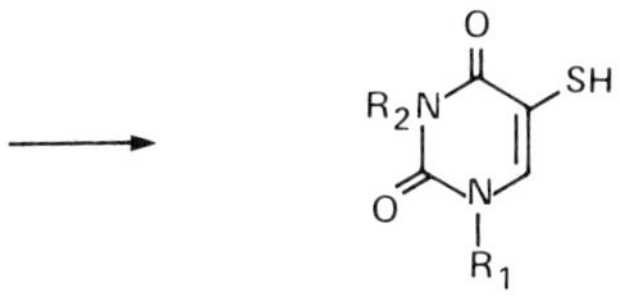

1, R_1 = β-D-ribofuranosyl; R_2 = H
2, R_1 = 2'-deoxy-β-D-ribofuranosyl; R_2 = H
14, R_1 = β-D-arabinofuranosyl; R_2 = H

Scheme 1. Preparation of 5-thiocyanopyrimidine nucleosides and their conversion to 5-mercaptopyrimidine nucleosides.

uridine, whereas Prystas and Sorm [34] obtained a 29% yield of 5-cyanouridine via a Hilbert-Johnson type synthesis. The second approach has relied upon efforts to transform a preformed nucleoside. Inoue and Ueda [35] and Ueda et al. [36] obtained the 2',3'-isopropylidene-5'-acetyl-5-cyanouridine by reacting the protected 5-bromouridine derivative with sodium cyanide. The thymidine analog, 5-cyano-2'-deoxyuridine has been prepared by treatment of a persilylated derivative of 5-bromo-2'-deoxyuridine with CuCN [37] or by reaction of the 3',5'-diacetate of 5-bromo-2'-deoxyuridine with KCN in DMSO [38].

5-Nitropyrimidine nucleosides

Wempen et al. [39] first reported the preparation of 5-nitrouridine by the nitric acid nitration of 2',3',5'-tri-O-(3,5-dinitrobenzoyl)uridine. The ribonucleoside has also been prepared by the mercuric cyanide-nitromethane condensation procedure [40]. A similar nitration approach to 5-nitro-2'-deoxyuridine yielded only glycoside bond cleavage [41], a finding confirmed in this laboratory also. An anomeric mixture of the di-O-toluyl derivative of 5-nitro-2'-deoxyuridine was obtained in 1% yield when 5-nitrouracil-mercury was reacted with the protected 2'-deoxy-D-ribofuranosyl chloride [41]. Kluepfel et al. [42], utilizing *trans-N*-deoxy-ribolase from Lactobacillus, reported an enzymatic synthesis of 5-nitro-2'-deoxyuridine, but limited structure proof was presented.

Potent antiviral properties were claimed for the enzymatically prepared 5-nitro-2'-deoxyuridine [42]. To find a less equivocal synthesis of this nitronucleoside and to provide a more practical access to this class of nucleoside, we investigated [43] the nitration of nucleosides and nucleotides by nitronium tetrafluoroborate/sulfolane, a reagent introduced by Olah [44]. There exists one earlier report on the use of nitronium tetrafluoroborate for the nitration of nucleotides, but without characterization, the nitration product was reduced to the amino form [45].

Uracil, 1-methyluracil and 1,3-dimethyluracil were all smoothly converted by the NO_2BF_4/sulfolane reagent to the corresponding 5-nitro derivatives in 80–90% yield (Scheme 2). Treatment of uridine or 2'-deoxyuridine with NO_2BF_4/sulfolane under identical conditions, however, led to extensive glycoside bond fission and to 5-nitrouracil as the major product. Use of a variety of protecting groups for the sugar hydroxyls (including isopropylidene, acetate, 3,5-dinitrobenzoate or 5'-O-nitrate), or alteration of reaction conditions did not markedly improve the outcome of the reaction. Treatment of the nucleoside 5'-monophosphates, however, with NO_2BF_4/sulfolane gave good yields of the corresponding 5-nitropyrimidine nucleotides with some 3'-O-nitration resulting in the 3'-O,5-dinitro-nucleoside as a byproduct. Treatment of 5-nitro-2'-deoxyuridine 5'-monophosphate with bacterial alkaline phosphatase gave 5-nitro-2'-deoxyuridine, the spectral properties of which were identical with the enzymatically prepared nucleoside. This represented the first successful approach to 5-nitro-2'-deoxyuridine. The side-product of the reaction of 2'-deoxyuridine 5'-monophosphate with NO_2BF_4 was 3'-O,5-dinitro-2'-deoxyuridine 5'-monophosphate which could be dephosphorylated to 3'-O,5-dinitro-2'-deoxyuridine. Similarly, uridine 5'-monophosphate gave 5-nitrouridine 5'-monophosphate as well as the 3'-O,5-dinitrouridine 5'-monophosphate.

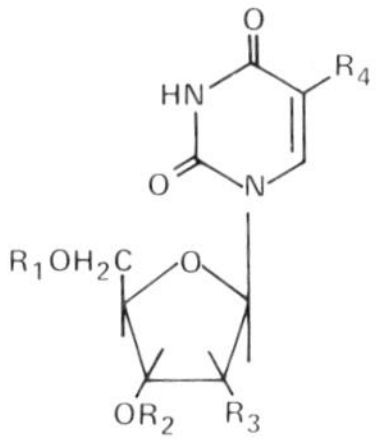

$R_1 = R_2 = R_4 = H$; $R_3 = OH$
$R_1 = R_2 = R_3 = R_4 = H$
$R_1 = R_2 = H$; $R_4 = NO_2$; $R_3 = OH$

$R_1 = HO-\overset{O}{\underset{OH}{P}}-$; $R_2 = R_4 = H$; $R_3 = OH$

$R_1 = HO-\overset{O}{\underset{OH}{P}}-$; $R_2 = H$; $R_3 = OH$; $R_4 = NO_2$

$R_1 = HO-\overset{O}{\underset{OH}{P}}-$; $R_2 = R_3 = R_4 = H$

$R_1 = HO-\overset{O}{\underset{OH}{P}}-$; $R_2 = R_3 = H$; $R_4 = NO_2$

$R_1 = HO-\overset{O}{\underset{OH}{P}}-$; $R_2 = R_4 = NO_2$; $R_3 = H$

$R_1 = R_2 = R_3 = H$; $R_4 = NO_2$
$R_1 = R_3 = H$; $R_2 = R_4 = NO_2$

Scheme 2.

5-O-Alkylated derivatives of 5-hydroxy-2'-deoxyuridine

5-Hydroxyuridine, in the presence of base, may be selectively alkylated on the pyrimidine 5-hydroxyl by allyl and propargyl halides [46,47]. We have extended this reaction to the synthesis of a variety of 5-O-alkylated derivatives of 5-hydroxy-2'-deoxyuridine [48]. Either 5-hydroxyuridine or 5-hydroxy-2'-deoxyuridine were treated with one equivalent of NaOH to give the monoanion at the 5-hydroxyl function and the O-alkylated by 1.5—2.0 equivalents of a properly

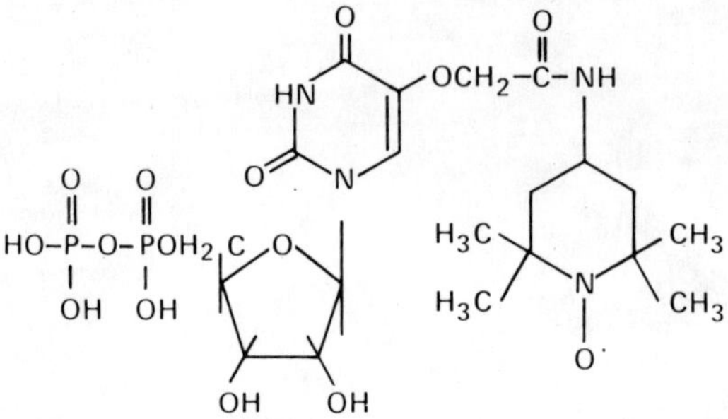

Scheme 3.

substituted activated allyl halide to give the corresponding 5-O-alkylated nucleoside in a yield of 40–60%. This approach has been successfully used to prepare the first site-specific spin-labeled nucleic acid [49]. Reaction of 5-hydroxyuridine-5'-diphosphate with 4-(α-chloroacetamido)-2,2,6,6-tetramethylpiperidino-1-oxy in the presence of one equivalent of sodium hydroxide gave the corresponding 5-O-alkylated derivative of 5-hydroxyuridine 5'-diphosphate (Scheme 3)

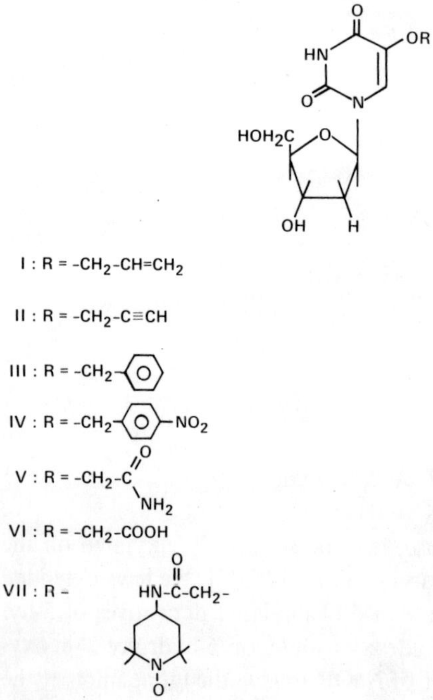

I : R = -CH$_2$-CH=CH$_2$

II : R = -CH$_2$-C≡CH

III : R = -CH$_2$-⟨O⟩

IV : R = -CH$_2$-⟨O⟩-NO$_2$

V : R = -CH$_2$-C(=O)NH$_2$

VI : R = -CH$_2$-COOH

VII : R = HN-C(=O)-CH$_2$- (tetramethylpiperidine-N-oxyl)

Scheme 4.

which could be copolymerized with UDP to give a spin-labeled polynucleotide.
Scheme 4 lists the derivatives of 2'-deoxyuridine that have been prepared by the above method.

Arabinosyl nucleosides

Based on the antiviral activity (vide infra) contingent on the introduction of certain substituents into position 5 of 2'-deoxyuridine, we have also prepared several new pyrimidine arabinosides. Two different approaches were employed. First uracil arabinoside was directly converted to the corresponding analog (Scheme 5) [19,50].

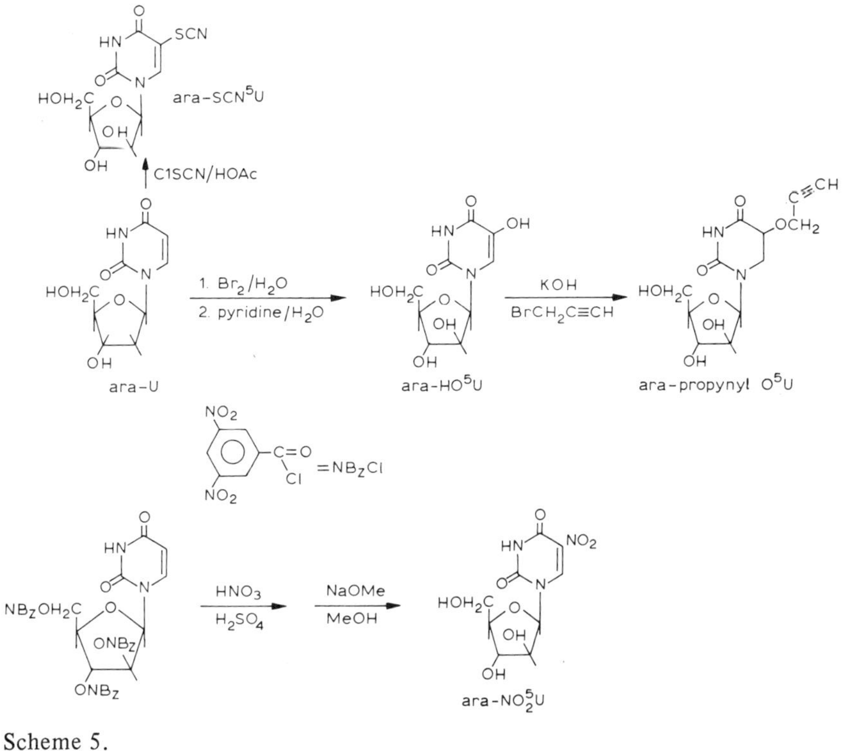

Scheme 5.

The second approach involved inversion of configuration at C–2 (Scheme 6) [50].

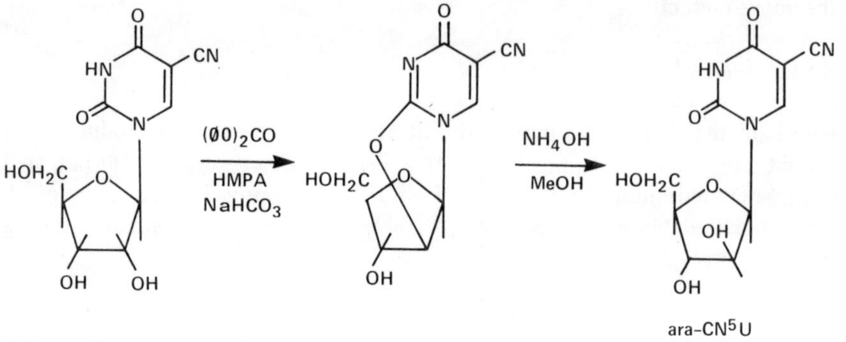

Scheme 6.

Antiviral and antimetabolic properties of 5-substituted 2'-deoxyuridine

Various 5-substituted 2'-deoxyuridine derivatives were investigated for their antiviral and antimetabolic properties in primary rabbit kidney (PRK) cells or human skin fibroblasts (HSF). These 2'-deoxyuridine derivatives could be divided into different classes depending on the nature of the 5-substituent:
(1) *halogeno atoms* ($X-C_5$), in which X represents fluorine, chlorine, bromine or iodine;
(2) *C-substituents* ($C-C_5$), if the atom attached to C_5 is carbon. Representative examples of such substituents are cyano, trifluoromethyl and hydroxymethyl;
(3) *N-substituents* ($N-C_5$), if the atom attached to C_5 is nitrogen. Examples of this group are amino and nitro.
(4) *O-substituents* ($O-C_5$), if the atom attached to C_5 is oxygen. Examples of this category are hydroxy, allyloxy, propynyloxy, benzyloxy, paranitrobenzyloxy and carboxamidomethoxy.
(5) *S-substituents* ($S-C_5$), if the atom attached to C_5 is sulfur. A representative example of this class is the thiocyano group.

The antiviral potential of 5-fluoro, 5-chloro, 5-bromo, 5-iodo and 5-trifluoromethyl-2'-deoxyuridines have been dealt with in recent reviews [51,52]. In contrast to earlier reports on the lack of anti-herpes and/or -vaccinia activity in a number of in vitro systems [53–56], 5-fluoro-2'-deoxyuridine proved, in this study, just as effective as the other halogeno-deoxynucleosides in protecting PRK cells from the cytopathic effect of vaccinia and herpes simplex viruses (Table 2).

Table 2. 5-Substituted 2'-deoxyuridine derivatives: effect on virus-induced cytopathogenicity in PRK cells.

Compound	ID_{50}	
	Vaccinia virus	Herpes simplex virus (Type 1)
5-fluoro-2'-deoxyuridine	0.1	0.1
5-chloro-2'-deoxyuridine	0.2	0.2
5-bromo-2'-deoxyuridine	0.1	0.1
5-iodo-2'-deoxyuridine	0.2	0.2
5-cyano-2'-deoxyuridine	4	40
5-amino-2'-deoxyuridine	1	10
5-nitro-2'-deoxyuridine	0.1	2
5-hydroxy-2'-deoxyuridine	4	4
5-hydroxymethyl-2'-deoxyuridine	4	2
5-tirfluoromethyl-2'-deoxyuridine	0.1	0.2
5-thiocyano-2'-deoxyuridine	4	4
5-allyloxy-2'-deoxyuridine	20	10
5-propynyloxy-2'-deoxyuridine	10	0.4
5-benzyloxy-2'-deoxyuridine	>100	40
5-para-nitrobenzyloxy-2'-deoxyuridine	>100	>100
5-carboxamidomethoxy-2'-deoxyuridine	40	20

ID_{50} = inhibitory dose-50: concentration (μg/ml) required to reduce virus-induced cytopathogenicity in PRK cells by 50%.

Compounds, such as 5-cyano and 5-amino-2'-deoxyuridine, inhibited vaccinia-induced cytopathogenicity at a concentration which was 10-fold lower than that required to inhibit herpes-virus-induced CPE. The ID_{50} for 5-nitro-2'-deoxyuridine was 10- to 100-fold lower than that needed to suppress herpes CPE. 5-Propynyloxy-2'-deoxyuridine behaved in the reverse manner, since it blocked herpes-induced CPE at a 25-fold lower concentration that that necessary for inhibition of vaccinia CPE. Other analogs were equally effective against herpes and vaccinia virus: e.g. 5-hydroxy-, 5-hydroxymethyl-, 5-thiocyano-, 5-allyloxy- and 5-carboxamidomethoxy-2'-deoxyuridines (Table 3).

The data in Table 2 showed that for the class of O-substituents examined, any alteration of the propynyloxy (propargyloxy) side chain had a negative effect on antiviral activity. When the triple bond of 5-propynyloxy-2'-deoxyuridine was replaced by the bulky phenyl or p-nitrophenyl groups, the resulting analogs were devoid of significant antiviral activity. Substitution of the acetylene by an olefinic function gave 5-allyloxy-2'-deoxyuridine which was 10- to 20-fold less

Table 3. Ratio of anti-herpes to anti-vaccinia activity of 2'-deoxyribosyl and arabinosyl nucleosides in PRK cells.

Compound	Ratio of ID_{50} (vaccinia virus) to ID_{50} (herpes simplex virus)
5-fluoro-2'-deoxyuridine	1
5-chloro-2'-deoxyuridine	1
5-bromo-2'-deoxyuridine	1
5-iodo-2'-deoxyuridine	2
5-cyano-2'-deoxyuridine	0.1
5-amino-2'-deoxyuridine	0.1
5-nitro-2'-deoxyuridine	0.1
5-hydroxy-2'-deoxyuridine	1
5-hydroxymethyl-2'-deoxyuridine	2
5-trifluoromethyl-2'-deoxyuridine	0.5
5-thiocyano-2'-deoxyuridine	1
5-allyloxy-2'-deoxyuridine	2
5-propynyloxy-2'-deoxyuridine	25
5-benzoyloxy-2'-deoxyuridine	2.5
5-para-nitrobenzyloxy-2'-deoxyuridine	–
5-carboxamidomethoxy-2'-deoxyuridine	2
adenine arabinoside	0.1
cytosine arabinoside	1
thymine arabinoside	10
uracil arabinoside	20

Ratios were calculated from individual ID_{50} values listed in Tables 6 and 10.

active against herpes. When the carboxamide group replaced the triple-bond of 5-propynyloxy-2'-deoxyuridine, there was also a significant drop in antiviral activity. Finally, all antiviral activity was abolished when the carboxyl group (negatively charged at physiological pH) replaced the acetylene group of 5-propynyloxy-2'-deoxyuridine.

On the basis of their vaccinia-ID_{50} to herpes-ID_{50} ratio (Table 3), 5-nitro-2'-deoxyuridine, 5-cyano-2'-deoxyuridine and 5-amino-2'-deoxyuridine could be considered *specific anti-vaccinia agents*. In the same way, 5-propynyloxy-2'-deoxyuridine could be considered a specific *anti-herpes agent*. Other 5-substituted uracil 2'-deoxyribosides have been accredited with specific anti-herpes activities. They include 5-propyl-2'-deoxyuridine [57], 5-methylamino-2'-deoxyuridine [56–58], 5-methoxymethyl-2'-deoxyuridine [59–61] and 5-mercaptomethyl-2'-deoxyuridine [62]. 5-Vinyl- and 5-allyl-2'-deoxyuridine [63] recently have been shown to inhibit herpes simplex replication, but it was not determined how the anti-herpes activity of these compounds compared with their other antiviral properties.

Addition of 2'-deoxythymidine (TdR), which by itself shows no antiviral activity, effectively reversed the antiviral activity of all TdR analogs (Table 4). Significant differences were noted in the reversing capacity of TdR, depending on the nature of the analog evaluated; e.g. 5-cyano-, 5-thiocyano-, 5-hydroxy- and 5-fluoro-2'-deoxyuridines lost their anti-herpes activity in the presence of TdR concentrations which were several orders of magnitude lower than the inhibitory concentrations of the analogs. To abolish the anti-herpes activity of compounds such as 5-iodo- or 5-bromo-2'-deoxyuridine, much higher TdR doses were required. The differences in the reversing potency of TdR most probably reflect differences in the mode of action of the various deoxythymidine analogs. 5-Iodo- and 5-bromo-2'-deoxyuridine interfere with a very late step in thymidine metabolism; that is, subsequent to incorporation into DNA. Thus, their antiviral action is not as readily reversed by TdR as the antiviral action of 5-fluoro-, 5-cyano- and 5-thiocyano-2'-deoxyuridine, which may act at a much earlier stage of thymidine metabolism.

Table 4. Reversal of antiviral activities of 5-substituted 2'-deoxyuridine derivatives by deoxythymidine.

Compound	ID_{100}	RD_{50}	Reversal ratio
5-fluoro-2'-deoxyuridine	1	0.004	250
5-chloro-2'-deoxyuridine	2	0.02	100
5-bromo-2'-deoxyuridine	1	0.3	3
5-iodo-2'-deoxyuridine	2	1	2
5-cyano-2'-deoxyuridine	400	0.1	4000
5-hydroxy-2'-deoxyuridine	40	0.1	400
5-hydroxymethyl-2'-deoxyuridine	20	0.1	200
5-trifluoromethyl-2'-deoxyuridine	2	2	1
5-thiocyano-2'-deoxyuridine	40	<0.1	>400
5-allyloxy-2'-deoxyuridine	100	0.4	250
5-propynyloxy-2'-deoxyuridine	10	0.07	150
5-carboxamidomethoxy-2'-deoxyuridine	200	<20	>10

ID_{100} = inhibiting dose$_{100}$ = concentration (μg/ml) required to reduce cytopathogenicity caused by herpes simplex virus (Type 1) in PRK cells by 100%. ID_{100} is generally $ID_{50} \times 10$.
RD_{50} = reversing dose$_{50}$ = deoxythymidine concentration (μg/ml) required to reduce protective activity of each compound (tested at 1 ID_{100}) by 50%.
Reversal ratio = ratio of ID_{100} to RD_{50}.

Do the 5-substituted 2'-deoxyuridines (Table 2) specifically interfere with viral nucleic acid metabolism, or is their antiviral activity merely the consequence of inhibition of host cell DNA biosynthesis? To address this question, uninfected PRK cells, which had been exposed to the analogs in question, were monitored by measurement of DNA synthesis, with either [^{3}H-methyl]-deoxythymidine or [2-^{14}C]-deoxyuridine as labeled probes. Most compounds, e.g. 5-chloro-, 5-bromo-, and 5-iodo-2'-deoxyuridine, inhibited [2-^{14}C]-dU and [^{3}H-methyl]-dT incorporation to the same extent (Table 5). Also, most compounds, including 5-fluoro-, 5-chloro-, 5-bromo-, 5-hydroxy-, 5-hydroxymethyl- and 5-allyloxy-2'-deoxyuridine and the antiviral drugs IUdR (5-iodo-2'-deoxyuridine) and F$_3$TdR (5-trifluoromethyl-2'-deoxyuridine) inhibited cellular DNA synthesis at similar or slightly higher concentrations than those required to block vaccinia or herpes simplex virus replication (compare Tables 2 and 5). However, 5-cyano-, 5-thiocyano- and 5-propynyloxy-2'-deoxyuridine were found to block

Table 5. 5-Substituted 2'-deoxyuridine derivatives: effect on PRK cell DNA synthesis as monitored by [^{3}H-methyl]deoxythymidine incoporation and [2-^{14}C]-deoxyuridine incorporation, respectively.

Compound	ID$_{50}$	ID$_{50}$
	[^{3}H-methyl]-deoxythymidine	[2-^{14}C]-deoxyuridine
5-fluoro-2'-deoxyuridine	100	0.0005
5-chloro-2'-deoxyuridine	0.3	0.2
5-bromo-2'-deoxyuridine	0.3	0.3
5-iodo-2'-deoxyuridine	2.5	1.2
5-cyano-2'-deoxyuridine	>400	75
5-nitro-2'-deoxyuridine	>100	0.2
5-hydroxy-2'-deoxyuridine	25	8
5-hydroxymethyl-2'-deoxyuridine	10	15
5-trifluoromethyl-2'-deoxyuridine	25	0.05
5-thiocyano-2'-deoxyuridine	>400	40
5-allyloxy-2'-deoxyuridine	50	45
5-propynyloxy-2'-deoxyuridine	75	50
5-benzyloxy-2'-deoxyuridine	50	NT
5-para-nitrobenzyloxy-2'-deoxyuridine	20	NT
5-carboxamidomethoxy-2'-deoxyuridine	300	150

ID$_{50}$ = inhibiting dose$_{50}$ = concentration (μg/ml) inhibiting [^{3}H-methyl]deoxythymidine or [2-^{14}C]deoxyuridine incorporation into cellular DNA by 50%. NT = not tested.

cellular DNA synthesis, as monitored by either [^{3}H-methyl]-dT or [2-^{14}C]-dU incorporation, at concentrations which were far in excess of those inhibiting vaccinia or herpes multiplication.

For 5-fluoro-, 5-trifluoromethyl-, 5-thiocyano-, 5-nitro- and 5-cyano-2'-deoxyuridine, the ID_{50} for inhibiting [2-^{14}C]-dU incorporation was significantly lower than the ID_{50} for inhibiting [^{3}H-methyl]-dT incorporation (Table 5). How might such analogs inhibit deoxyuridine incorporation while leaving deoxythymidine incorporation unaffected? Is there some difference between the metabolic pathways leading to dU incorporation into DNA, on the one hand, and dT incorporation, on the other? As illustrated in Scheme 7, upon introduction into the cell, dT and DU are both converted by deoxythymidine kinase to their respective 5'-monophosphate, dTMP and dUMP. dTMP is then processed

METABOLIC PATHWAYS LEADING TO INCORPORATION
OF DEOXYTHYMIDINE (TdR) AND DEOXYURIDINE (UdR)
INTO DNA

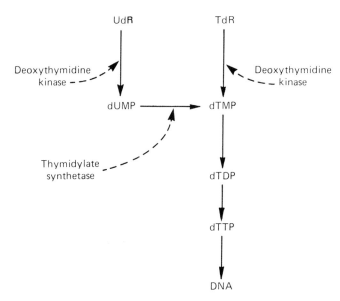

Scheme 7.

to dTDP, dTTP and finally incorporated into DNA. To be incorporated into DNA, dUMP needs an additional step, namely, methylation at pyrimidine C_5 by thymidylate synthetase. Thus, 2'-deoxyuridine analogs which block dU but not dT incorporation into DNA may be expected to operate at the thymidylate synthetase level. To do so, these analogs must, of course, be phosphorylated to their 5'-monophosphates. Both 5-fluoro- and 5-trifluoromethyl-2'-deoxyuridine previously have been shown to inhibit thymidylate synthetase in vitro [64]. Thus, the data presented in Table 5 extend the reported in vitro results to an in vivo system and further imply that 5-cyano-, 5-nitro- and 5-thiocyano-2'-deoxyuridine, in analogy with 5-fluoro- and 5-trifluoromethyl-2'-deoxyuridine, may specifically block the dUMP → dTMP conversion catalyzed by thymidylate synthetase.

A further point worth consideration is the behavior of the thiocyano nucleosides as antiviral agents. The facile reduction of such thiocyano nucleosides to the 5-mercapto derivatives has already been mentioned. The question arises as to whether the antiviral activity of 5-thiocyano-2'-deoxyuridine is due to its in vivo conversion to 5-mercapto-2'-deoxyuridine or due to the pseudohalogen nature of the intact thiocyano group analogous to the halogenodeoxyuridines. That the former possibility may be the case is suggested by the above labeling studies, coupled with the ease of reversal of its antiviral activity by dT. Both observations imply thymidylate synthetase as a primary target of activity, and 5-mercapto-2'-deoxyuridine has already been established as a potent inhibitor of the synthetase [31]. Alternatively, the thiocyanonucleoside may itself block thymidylate synthetase. Another possible mode of action of thiocyano nucleosides is suggested by the observation that the tri-O-acetate of 5-thiocyanouridine, a riboside, is inhibitory toward vesicular stomatitis virus [65], an RNA virus. This triacetate was shown directly to inactivate VSV extracellularly. It is possible that this effect may result from the reduction of the thiocyano group. This would occur by (non-enzymatic) transfer of the cyano group from the nucleoside to a mercapto group, possibly to an enzyme required for virus replication:

$$RSCN + R'SH \rightarrow RSH + R'SCN$$

The fact that the triacetate alone, and not the unblocked riboside or deoxyriboside, is active against VSV could be rationalized by the increased hydrophobicity of the acetate, resulting in greater membrane permeability. A considerable number of nucleoside analogs have been evaluated for their effects on parasites; namely, Schistosoma mansoni and Brugia pahangi [66]. In an in vitro test system *both* 5-thiocyanouridine and 5-thiocyano-2'-deoxyuridine caused the death of the parasites at 5×10^{-5} M, whereas no other nucleoside evaluated, including IDU and FUdR, had any effect, even at 10^{-4} M. These results also suggest an action that may not be directly related to nucleic acid metabolism.

Arabinosyl nucleosides

Although known for some time, ara-U and ara-T have been evaluated only cursorily for their antiviral potentials. Ara-U was found ineffective against herpes simplex, vaccinia, measles and polio virus [67]. Ara-T, although as effective an anti-herpes and anti-vaccinia agent as ara-C and ara-A in cell culture [67,68], showed only modest activity on the herpes-infected rabbit eye [68]. Recent evidence points to some specificity in the anti-herpes activity of ara-T [69,81].

The antiviral activity of ara-U and ara-T has now been re-evaluated, and, surprisingly, ara-U was found to exert a distinct inhibitory effect on the cytopathogenicity induced by herpes simplex (Table 6). Of all arabinosyl nucleosides tested, ara-U exhibited the highest vaccinia ID_{50} to herpes ID_{50} ratio viz. 20. The vaccinia ID_{50} to herpes ID_{50} ratio was 10 for ara-T, 1 for ara-C and only 0.1 for ara-A. Accordingly, ara-U and ara-T could be considered as *specific antiherpes agents*. In contrast to ara-C and ara-A, neither ara-U nor ara-T inhibited cellular DNA synthesis, as monitored by either [^{3}H-methyl]-dT or [2-^{14}C]-dU incorporation, unless extraordinarily high doses (300 µg/ml) were employed. Ara-C and ara-A blocked [^{3}H-methyl]-dT incorporation and [2-^{14}C]-dU incor-

Table 6. Arabinosyl nucleosides: effect on virus-induced cytopathogenicity in PRK cells.

Compound	ID_{50}	
	vaccinia virus	herpes simplex virus (Type 1)
adenine arabinoside	0.4	4
cytosine arabinoside	0.04	0.04
thymine arabinoside	1	0.1
uracil arabinoside	200	10
5-cyanouracil arabinoside	>100	>100
5-hydroxyuracil arabinoside	>200	>100
5-propynyloxyuracil arabinoside	>200	20
5-nitrouracil arabinoside	>200	>200
5-thiocyanouracil arabinoside	40	4

ID_{50} = inhibitory dose$_{50}$: concentration (µg/ml) required to reduce virus-induced cytopathogenicity by 50%.

poration to the same extent. The concentrations at which ara-C inhibited vaccinia and herpes simplex virus replication (Table 6) corresponded closely to those found to inhibit [^{3}H-methyl]-dT and [2-^{14}C]-dU incorporation into host cell DNA. Hence, ara-C cannot be considered as a selective antiviral agent.

The pyrimidine C-5 modifications which imparted potent antiviral activity to 2'-deoxyuridine failed to give nucleoside arabinosides with significant antiviral (Table 6) or antimetabolic activity (Table 7). Thus, substitution of cyano, propynyloxy, hydroxy or nitro at C-5 of uracil arabinosides abolished the activity associated with ara-U. The only exception to this was 5-thiocyanouracil arabinoside which could be considered a selective anti-herpes agent (Table 6).

Basis for specific antiviral activity

The data reviewed in Tables 2, 3, 5, 6, and 7 reveal a number of nucleoside analogs which inhibit vaccinia and/or herpes simplex virus replication at a concentration far below the concentration inhibiting normal cellular DNA biosynthesis. How could the selective anti-herpes activity of ara-U, ara-T, and 5-propynyloxy-2'-deoxyuridine and the anti-vaccinia activity of 5-cyano-, and 5-amino-2'-deoxyuridine be rationalized? We may postulate that, wherever viral nucleic acid metabolism diverges from normal cellular RNA or DNA metabolism, either quantitatively (e.g. in the amount of purine or pyrimidine precursors required for RNA or DNA synthesis), or qualitatively (e.g. in the substrate specificity of the enzymes involved in nucleic acid biosynthesis), viral nucleic

Table 7. Arabinosyl nucleosides: effect on PRK cell DNA synthesis as monitored by [^{3}H-methyl]-deoxythymidine incorporation and [2-^{14}C]deoxyuridine incorporation, respectively.

Compound	ID_{50} [^{3}H-methyl]-deoxythymidine	ID_{50} [2-^{14}C]-deoxyuridine
adenine arabinoside	25	20
cytosine arabinoside	0.1	0.05
thymine arabinoside	300	100
uracil arabinoside	300	400
5-cyanouracil arabinoside	>100	>100
5-hydroxyuracil arabinoside	>200	>200
5-propynyloxyuracil arabinoside	>200	>200
5-nitrouracil arabinoside	>200	>200

ID_{50} = inhibiting dose$_{50}$: concentration (µg/ml) inhibiting [^{3}H-methyl]- or [2-^{14}C]-deoxyuridine incorporation into cellular DNA by 50%.

acid metabolism would become particularly vulnerable to chemotherapeutic attack.

Vaccinia virus and herpes simplex virus offer some good targets for a selective chemotherapy. Once they have invaded the cell, they code for a number of enzymes (e.g. DNA polymerase, deoxythymidine kinase, deoxycytidine kinase, exodeoxyribonuclease, endodeoxyribonuclease etc.) which differ from the corresponding host-coded enzymes in such properties as heat stability, pH optimum, Michaelis constant (K_m) and antigenicity. One of the best characterized virus-coded enzymes is the pyrimidine deoxyribonucleoside kinase induced by herpes and simplex virus [2–8,70]. This enzyme phosphorylates both deoxythymidine and deoxycytidine. Both enzyme activities have a lower pH optimum and higher heat stability than the respective host enzymes, and their K_m is lower than that of the host enzymes [6,7]. Thus, the virus-induced enzymes have higher affinity for their substrates than the corresponding host enzymes. Mutant cells lacking TdR kinase activity have been irreversibly transformed by herpes simplex virus to yield cells which actively produce TdR kinase, and the enzyme expressed in these cells behaves antigenically as a virus-specified enzyme [4,8].

Due to the broadened substrate specificity of the herpes-virus-coded pyrimidine deoxyribonucleoside kinase as compared to the corresponding host enzyme, some pyrimidine deoxyribonucleoside analogs which do not serve as substrates for the cellular kinases, hence are not phosphorylated in normal cells, may be phosphorylated in herpes simplex virus-infected cells. These compounds may then be expected to block selectively the replication of herpes simplex virus. Such a modus operandi appears to account for the selective anti-herpes activity of 5-bromo- and 5-iodo-2'-deoxycytidine [71,72,82].

If arabinosyl nucleosides, such as ara-U, and 5-substituted 2'-deoxyuridines, such as 5-propyl-2'-deoxyuridine, need to be phosphorylated by the herpes-virus-induced deoxythymidine kinase in order to be effective as anti-herpes agents, one might expect herpes virus strains which do not induce such an enzyme not to be sensitive to the inhibitory effects of ara-U, 5-propyl-2'-deoxyuridine and their congeners. This expectation has been fulfilled. Herpes simplex virus (Type 2) strain 333 did not induce deoxythymidine kinase in either PRK or HSF cells (Fig. 1) [73]. Concomitantly, the replication of this virus strain was not inhibited by any of the pyrimidine nucleosides tested (Table 8). When assayed under identical conditions, with the same multiplicity of infections, herpes simplex (Type 1) strain KOS effectively induced TdR kinase activity in both PRK and HSF cells (Fig. 1) and, concomitantly, the replication of herpes strain KOS was inhibited by all deoxyuridines and arabinosides tested (Table 8). These data establish the necessity of herpes-induced TdR kinase activity for TdR analogs to be effective as anti-herpes agents.

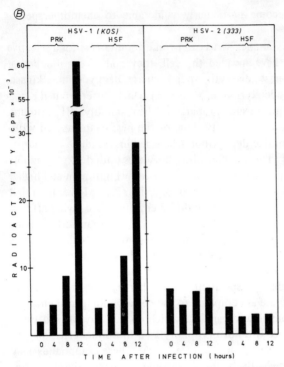

Fig. 1. Deoxythymidine kinase activities induced by herpes simplex virus strains KOS (Type 1) and 333 (Type 2) in primary rabbit kidney (PRK) cells and human skin fibroblast (HSF) cells.

If TdR analogs owe their anti-herpes activity to the conversion of the analog to its 5'-monophosphate by a virus-coded TdR kinase, one may also expect nucleosides which do not require the virus-induced TdR kinase to be equally effective against herpes strains which either induce or do not induce TdR kinase. This proved to be the case, for ara-A, the biologic activity of which depends on a cellular deoxyadenosine kinase [74], and which inhibited the replication of herpes simplex strain KOS and strain 333 to approximately the same extent (Table 8).

Whether or not a given nucleoside analog exerts a selective antiviral effect will, of course, depend on the interplay of both cell-coded and virus-coded enzymes, e.g. phosphorylases and kinases. In the uninfected cell, compounds such as ara-U, ara-T and 5'-propynyloxy-2'-deoxyuridine may undergo phos-

Table 8. Differential sensitivities of herpes simplex virus strains KOS (Type 1) and 333 (Type 2) to inhibitory effects of 2'-deoxyribosyl and arabinosyl nucleosides in PRK and HSF cells.

Compound	ID_{50}			
	HSV-1 (KOS)		HSV-2 (333)	
	PRK	HSF	PRK	HSF
5-fluoro-2'-deoxyuridine	0.4	0.1	>200	>200
5-chloro-2'-deoxyuridine	0.4	0.1	>200	>200
5-bromo-2'-deoxyuridine	0.4	0.1	>200	>200
5-iodo-2'-deoxyuridine	0.4	0.1	>200	>200
5-cyano-2'-deoxyuridine	20	150	>200	>200
5-hydroxy-2'-deoxyuridine	4	1	>200	>200
5-hydroxymethyl-2'-deoxyuridine	2	2	>200	>200
5-trifluoromethyl-2'-deoxyuridine	0.4	0.2	>200	>200
5-thiocyano-2'-deoxyuridine	20	4	>200	>200
5-allyloxy-2'-deoxyuridine	4	40	>200	>200
5-propynyloxy-2'-deoxyuridine	0.7	1	>200	>200
5-carboxamidomethoxy-2'-deoxyuridine	40	70	>200	>200
adenine arabinoside	4	1	10	10
cytosine arabinoside	0.07	0.02	150	>200
thymine arabinoside	0.4	0.1	>200	>200
uracil arabinoside	10	7	>200	>200

ID_{50} = inhibitory dose$_{50}$ = concentration (μg/ml) required to reduce virus-induced cytopathogenicity in PRK or HSF cells by 50%.

phorolytic cleavage before they can act as substrates for the cellular kinases. In herpes-virus-infected cells, however, the same compounds may be readily processed to their 5'-phosphates by the virus-induced deoxythymidine kinase before they are cleaved by nucleoside phosphorylases.

Provided that 5-propynyloxy-2'-deoxyuridine and the other TdR analogs are converted to their respective 5'-monophosphates by the herpes-coded TdR kinase, how would these compounds eventually arrest the replication of herpes simplex virus? As shown in Scheme 8, different targets may be envisaged, including the thymidylate synthetase, which converts dUMP to dTMP, or the kinases which convert dTMP to dTDP and finally dTTP. If the 2'-deoxyurdine-5'-monophosphates are phosphorylated intracellularly to the 5'-triphosphates, they may interfere with the DNA polymerase reaction. Some of the 2'-deoxyuridine derivatives may even be incorporated into DNA and, in analogy to IUdR and

POSSIBLE POINTS OF ATTACK OF 5-SUBSTITUTED 2'-DEOXYURIDINE DERIVATIVES (5-X-UdR) ON HERPES SIMPLEX VIRUS REPLICATION

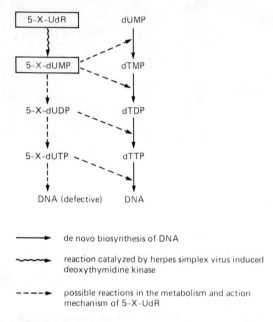

Scheme 8.

BrUdR, impair DNA replication and RNA transcription. Whatever the exact mode of action of our newly developed nucleoside analogs might be, if the inhibitory effect on herpes simplex virus is indeed the result of a more efficient phosphorylation of the nucleoside to its 5'-monophosphate in the virus-infected cell, the analog should prove specifically effective for virus-directed DNA synthesis and not (or much less) interfere with normal cellular DNA synthesis.

Cytotoxicity of 5-substituted pyridimine deoxyribonucleosides

For most of the nucleosides listed in Table 2, antiviral indexes were determined. The antiviral index was defined as the minimum toxic dose (required to reduce the multiplication of exponentially growing PRK cells by 30%) divided by the minimum effective dose (required to reduce virus replication in stationary PRK cell cultures by 50%). Table 9 shows that 5-fluoro- and 5-hydroxymethyl-2′-deoxyuridine were cytostatic at doses which were 2- to 10-fold lower than the doses which proved inhibitory to vaccinia or herpes simplex virus. The established antiviral agents, IUdR and F_3TdR, showed a modest antiviral index (10–20). The highest safety margin was displayed by 5-propynyloxy-2′-deoxyuridine that did not impair normal cell growth at concentrations as high as 100–300 μg/ml.

Summary and conclusions. The new insights emerging from this report could be summarized as follows:
(1) ara-U, a compound which has been considered pharmacologically inert in the past, possesses specific anti-herpes activity (in vitro);
(2) both 5-cyano-2′-deoxyuridine [*38*] and 5-amino-2′-deoxyuridine are distinguished by specific anti-vaccinia activity;
(3) 5-propynyloxy-2′-deoxyuridine which inhibited herpes-virus-induced cytopathogenicity at a concentration 40-fold lower than that required to inhibit vaccinia virus [*48*] is, therefore, selective in its antiviral action:
(4) 5-nitro-2′-deoxyuridine and its 5′-monophosphate [*75*] possess potent antiviral activity;

Table 9. 5-Substituted 2′-deoxyuridine derivatives: antiviral indexes in PRK cells.

Compound	TD_{30}	Antiviral index [a,b]
5-fluoro-2′-deoxyuridine	0.01	0.1 (b, c)
5-bromo-2′-deoxyuridine	2	20 (b, c)
5-iodo-2′-deoxyuridine	4	20 (b, c)
5-trifluoromethyl-2′-deoxyuridine	1	10 (b)
5-cyano-2′-deoxyuridine	40	10 (b)
5-nitro-2′-deoxyuridine	3	15 (b)
5-thiocyano-2′-deoxyuridine	100	25 (b, c)
5-hydroxymethyl-2′-deoxyuridine	1	0.5
5-propynyloxy-2′-deoxyuridine	>100	250 (c)

TD_{30} = toxic dose$_{30}$ (μg/ml) required to inhibit cell growth by approximately 30% (determined by the coulter counter after 3 days incubation of exponentially growing PRK cells with varying concentrations of the compound). Input: 10^6 cells per petri dish. Output (control): about 6×10^6 cells per petri dish.
[a,b] Determined by dividing TD_{30} by ID_{50}. The minimum effective dose corresponds to the dose required to reduce the cytopathogenicity of vaccinia and/or herpes simplex by 50% (see Table 2). a: antivaccinia index; b: antiherpes index.

(5) several nucleosides, such as 5-nitro-2'-deoxyuridine, 5-cyano-2'-deoxyuridine and 5-thiocyano-2'-deoxyuridine act at the thymidylate synthetase level [75], as demonstrated by an in vivo assay; and
(6) the anti-herpes activity of various deoxythymidine analogs is stringently dependent on the presence of a specific virus-induced (deoxythymidine) kinase in the infected cell [73].

The studies reported here and from other laboratories have defined a number of new pyrimidine nucleosides which are prime candidates for further evaluation as antiviral drugs. These include (1) 5-ethyl-2'-deoxyuridine [9,78], which does not appear to be mutagenic and does not activate oncornavirus expression [77], although it is incorporated into DNA; (2) 5-cyano-2'-deoxyuridine [38], which is not incorporated into DNA at all [37], and which may specifically interact at the thymidylate synthetase level [76]; (3) 5-thiocyano-2'-deoxyuridine [19,65] and (4) 5-nitro-2'-deoxyuridine [75], which would also specifically block thymidylate synthetase [75]; (5) 5-methylamino-2'-deoxyuridine [56,58]; (6) 5-methoxymethyl-2'-deoxyuridine [59,60]; (7) 5-propyl-2'-deoxyuridine [57,63] and (8) 5-propynyloxy-2'-deoxyuridine [48], all of which could be considered as specific anti-herpes agents; and finally 5-iodo-5'-amino-2',5'-dideoxyuridine [11], also a specific anti-herpes agent, but with a novel mechanism of action [12,13].

These studies demonstrate that subtle chemical changes modulate subtle and differential responses to cell and virus replication with dramatic consequences in terms of selectivity and mechanisms of action. In doing so, they prove unquestionably the continuing potential of nucleoside analogs for antiviral chemotherapy.

References

1. Chan T.-S. (1977) Proc. Natl. Acad. Sci. USA, 74, 1734.
2. Dubbs D.R., Kit S. (1964) Virology, 22, 493.
3. Buchan A., Watson D.H., Dubbs D.R., Kit S. (1970) J. Virol., 5, 817.
4. Munyon W., Kraiselburd E., Davis D., Mann J. (1974) J. Virol., 7, 813, 140.
5. Davis D.B., Munyo W., Buchsbaum R., Chawda R. (1974) J. Virol., 24, 465.
6. Jamieson A.T., Gentry G.A., Subak-Sharpe J.H. (1974) J. Gen. Virol., 24, 465.
7. Jamieson A.T., Subak-Sharpe J.H. (1972) J. Gen. Virol., 24, 481.
8. Thoules M.E., Chadha K.C., Munyon W.H. (1976) Virology, 69, 350.
9. Swierskowski M., Shugar D. (1969) J. Med. Chem., 12, 533.
10. Shugar D. (1978) FEBS Lett., Supp., 40, S48.
11. Cheng Y.-C., Goz B., Neenan J.P., Ward D.C. and Prusoff W.H. (1975) J. Virol., 15, 1284.

12. Lin T.-S., Neenan J.P., Cheng Y.-C., Prusoff W.H. and Ward D.C. (1976) J. Med. Chem., *19*, 495.
13. Chen M.S., Ward D.C., Prusoff W.H. (1976) J. Biol. Chem., *251*, 4833.
14. Chen M.S., Ward D.C., Prusoff W.H. (1976) J. Biol. Chem., *251*, 4839.
15. Torrence P.F., Ledley G., unpublished observations.
16. Leo A., Hansch C., Elkins D. (1971) Chem. Rev., *71*, 525.
17. Berens K., Shugar D. (1963) Acta Biochim. Pol., *10*, 25.
18. Nagamachi T., Torrence P.F., Waters J.A., Witkop B. (1972) J. Chem. Soc. D, 1025.
19. Nagamachi T., Fourrey J.-L., Torrence P.F., Waters J.A., Witkop B. (1974) J. Med. Chem., *17*, 403.
20. Angus A.B., Bacon R.G.R. (1958) J. Chem. Soc., 744.
21. Guy R.G., Pearson I. (1973) J. Chem. Soc. Perkin Trans., 1, 281.
22. Bacon R.G.R., Guy R.G. (1958) J. Chem. Soc., 318.
23. Lutz W.B., Creveling C.R., Daly J.W., Witkop B., Goldberg L.I. (1972) J. Med. Chem., *15*, 795.
24. Bacon R.G.R. (1961) In: N. Kharasch, ed.: *Organic Sulfur Compounds*. Pergamon Press, New York, N.Y., p. 320.
25. Olson R.K., Snyder H.R. (1965) J. Org. Chem., *30*, 184.
26. Jocelyn P.C. (1972) *Biochemistry of the SH Group*. Academic Press, New York, N.Y.
27. Bardos T.J., Kotick M.P., Szantay C. (1966) Tetrahedron Lett., 1759.
28. Kotick M.P., Szantay C., Bardos I.J. (1969) J. Org. Chem., *34*, 3806.
29. Szekeres G.L., Bardos T.J. (1970) J. Med. Chem., *13*, 708.
30. Baranski K., Bardos T.J., Bloch A., Kalman T.I. (1969) Biochem. Pharmacol., *18*, 347.
31. Kalman T.I., Bardos T.J. (1970) Mol. Pharmacol., *6*, 621.
32. Shaw B., Warrener R.N., Maguire M.H., Ralph R.K. (1958) J. Chem. Soc., 2294.
33. Watanabe K., Fox J.J. (1969) J. Heterocycl. Chem., *6*, 109.
34. Prystas M., Sorm F. (1966) Collect. Czech. Chem. Commun., *31*, 3990.
35. Inoue H., Ueda T. (1971) Chem. Pharm. Bull., *19*, 1743.
36. Ueda T., Inoue H., Matsuda A. (1975) Ann. N.Y. Acad. Sci., *255*, 121–130.
37. Bleackley R.C., Jones A.S., Walker R.T. (1975) Nucleic Acids Res., *2*, 683.
38. Torrence P.F., Bhooshan B., Descamps J., De Clercq E. (1977) J. Med. Chem., *20*, 974.
39. Wempen I., Doerr I.L., Kaplan L., Fox J.J. (1966) J. Amer. Chem. Soc., *82*, 1624.
40. Watanabe K.A., Fox J.J. (1969) J. Heterocycl. Chem., *6*, 109.
41. Prystas M., Sorm F. (1965) Collect. Czech. Chem. Commun., *30*, 1900.
42. Kluepfel D., Murthy Y.K.S., Sartori G. (1965) Farmaco Ed. Sci., *20*, 757.
43. Huang G.-F., Torrence P.F. (1977) J. Org. Chem., *42*, 3821–3824.
44. Kuhn S.J., Olah G.A. (1961) J. Amer. Chem. Soc., *83*, 4564.
45. Shibaev V.N., Eliseeva H.I., Kochetkov N.K. (1972) Dokl. Akad. Nauk SSSR *203*, 860.
46. Otter B.A., Taube A., Fox J.J. (1971) J. Org. Chem., *36*, 1251.
47. Otter B.A., Saluja S.S., Fox J.J. (1977) J. Org. Chem., *37*, 2858.

48. Torrence P.F., Spencer J.W., Bobst A.M., Descamps J., De Clercq E. (1978) J. Med. Chem., 21, 228–231.
49. Bobst A.M., Torrence P.F. (1978) Polymery, 19, 115–117.
50. Torrence P.F., Huang G.-G., Edwards M., Bhooshan B., Descamps J., De Clercq E., J. Med. Chem., in press.
51. Ch'ien L.T., Schabel F.M., Jr., Alford C.A., Jr. (1973) In: W.A. Carter, ed.: Selective Inhibitors of Viral Functions. CRC Press, Cleveland, Ohio, p. 227.
52. Sugar J., Kaufman H.E. (1973) In: W.A. Carter, ed.: Selective Inhibitors of Viral Functions. CRC Press, Cleveland, Ohio, p. 295.
53. Herrmann E.C., Jr. (1961) Proc. Soc. Exp. Biol. Med., 107, 142.
54. Herrmann E.C., Jr. (1968) Appl. Microbiol., 16, 1151.
55. Kaufman H.E., Maloney E.D. (1963) Proc. Soc. Exp. Biol. Med., 112, 4.
56. Shen T.Y., McPherson J.F., Linn B.O. (1966) J. Med. Chem., 9, 366.
57. De Clercq E., Descamps J., Shugar D., to be published.
58. Nemes M.M., Hilleman M.R. (1965) Proc. Soc. Exp. Biol. Med., 119, 515.
59. Meldrum J.B., Gupta V.S., Saunders J.R. (1974) Antimicrob. Agents Chemother., 6, 393.
60. Babiuk L.A., Meldrum B., Gupta V.S., Rouse B.T. (1975) Antimicrob. Agents Chemother., 8, 643.
61. Babiuk L.A., Rouse B.T. (1975) Infect. Immun., 12, 1281.
62. Gupta V.S., Bubbar G.L., Meldrum J.B., Saunders J.R. (1975) J. Med. Chem., 18, 973–976.
63. Cheng Y.-C., Domin B.A., Sharma R.A., Bobek M. (1976) Antimicrob. Agents Chemother., 10, 119.
64. Reyes P., Heidelberger C. (1965) Mol. Pharmacol., 1, 14.
65. De Clercq E., Torrence P.F., Waters J.A., Witkop B. (1975) Biochem. Pharmacol., 24, 2171.
66. Jaffe J.J., Torrence P.F., unpublished observations.
67. De Ruder J., Privat de Garilhe M. (1966) Antimicrob. Agents Chemother., 1, 578.
68. Underwood G.E., Wisner C.A., Weed S.D. (1964) Arch. Ophthalmol., 72, 505.
69. Gentry G.A., Aswell J.F. (1975) Virology, 65, 294–296.
70. Cheng Y.-C., Ostrander M. (1976) J. Biol. Chem., 251, 2605.
71. Cooper G.M. (1973) Proc. Natl. Acad. Sci. USA, 70, 3788–3792.
72. Schildkraut I., Cooper G.M., Greer S. (1975) Mol. Pharmacol., 11, 153.
73. De Clercq E., Krajewska E., Descamps J., Torrence P.F. (1977) Mol. Pharmacol., 13, 980.
74. Krajewska E., De Clercq E., Shugar D., to be published.
75. De Clercq E., Descamps J., Huang G.-F., Torrence P.F. (1978) Mol. Pharmacol., 14, 422–430.
76. Guari K.K., Shif I., Wolford R.G. (1976) Biochem. Pharmacol., 25, 1809.
77. Preston C.M. (1977) J. Virol., 23, 455.
78. De Clercq E., Shugar D. (1975) Biochem. Pharmacol., 24, 1073.
79. De Clercq E., Luczak M., Shugar D., Torrence P.F., Waters J.A., Witkop B. (1975) Proc. Soc. Exp. Biol. Med., 151, 487.
80. Prusoff W.H., Ward D.C., Lin T.S., Chen M.S., Shain G.T., Chai D., Lentz E.,

Capizzi R., Idriss J., Ruddle N.H., Black F.L., Kumari H.L., Albert D., Bhatt P.N., Hsiung G.D., Strickland S., Cheng Y.-C. (1977) Ann. N.Y. Acad. Sci., *284*, 335.
81. Miller R.L., Iltis J.P., Rapp F. (1977) J. Virol., *23*, 679.
82. Doberson M.J., Jerkofsky M., Greer S. (1976) J. Virol., *20*, 478.

Yu.A. Ovchinnikov and M.N. Kolosov (eds.) Frontiers in Bioorganic Chemistry
and Molecular Biology © 1979, Elsevier/North-Holland Biomedical Press

CHAPTER 5

Structure and properties of boromycin and its degradation products

V. PRELOG

Abstract

Previous work on the structure of boromycin, of its degradation products and of aplasmomycin is reviewed and discussed. A hydrolysis product of boromycin, the desvalino-boromycin-anion is a negatively charged ionophore possessing high selectivity for D-enantiomers of phenylglycine derivatives.

Isolation and structure of boromycin

Several years ago a neutral lipophilic antibiotic was obtained from cultures of a Streptomyces strain isolated from soil collected at the Ivory Coast. This antibiotic was active, not only against gram-positive microorganisms, but also against Plasmodia (microorganisms causing malaria), which is rather unusual for a microbial metabolite. The Streptomyces strain was eventually identified as Streptomyces antibioticus [1]. The antibiotic left after combustion always a small amount of ash, which was identified as boric acid by atomic absorption and conversion into crystalline triethanolamine-ester.

Boric acid is ubiquitous in products of plant and animal origin (10–100 ppm) and even in drinking water (0.1–5 ppm). The boron content of the nutrition medium used was more than sufficient for the relatively low yield of the antibiotic (ca. 2 mg/l corresponding to about 25 μg B; the yield claimed in [1] is too high by a factor of 10). By changing the culture medium slightly and adding borate a 20-fold increase in yield of antibiotic (42 mg/l) could be obtained. Because of its boron content the antibiotic was given the name *boromycin*. For ten years it remained the only boron-containing organic natural compound. Only recently a new, closely related antibiotic was obtained by Y. Okami et al. [5] from cultures of a Streptomyces griseus strain isolated from a shallow sea mud. Its structure, determined by X-ray analysis by H. Nakamura et al. [6], will be discussed and compared with the structure of boromycin and its degradation products later in this paper.

The empirical formula $C_{44}H_{72}O_{15}NB$, first derived for boromycin from analytical results, had to be revised on the basis of the X-ray structural analyses of its derivatives to $C_{45}H_{74}O_{15}NB$. Spectroscopic data indicated that boromycin contains carbonyls, hydroxyls and a number of methyl groups. Further information about the constitution was obtained by degradation, as shown below. On

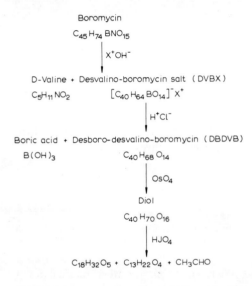

mild hydrolysis with sodium hydroxide, boromycin gives one equivalent of the (unnatural!) D-valine and a lipophilic sodium salt of an anion $C_{40}H_{64}O_{14}B^-$ that we shall call desvalino-boromycin-anion (DVB$^-$): on treatment of DVBNa with dilute hydrochloric acid one equivalent of boric acid is split off, and a neutral boron-free compound $C_{40}H_{68}O_{14}$, desboro-desvalino-boromycin (DBDVB) is obtained. The N- and B-free DBDVB seemed to be a convenient starting material for further degradation. After several unsuccessful attempts the following route was chosen. The double bond, identified by NMR, was oxidized by osmium(VIII)-oxide to a crystalline diol $C_{40}H_{70}O_{16}$, which, on further oxidation with periodic acid yielded three degradation products: (a) acetaldehyde, isolated as 2,4-dinitrophenylhydrazone, (b) an acid $C_{13}H_{22}O_4$, and (c) a neutral compound $C_{18}H_{32}O_5$. The two latter products, which were investigated extensively by

NMR, IR, and MS, showed many common structural features. The possibility that both compounds originate from the same part of the boromycin molecule could therefore not be excluded. It discouraged us from further degradation experiments and stimulated us to ask our colleagues to investigate a convenient derivative of boromycin by X-ray analysis.

To facilitate this task the caesium salt of DVB$^-$ was prepared by hydrolysis of boromycin with caesium hydroxide and submitted to X-ray analysis. Unfortunately, crystallization from warm aqueous methanol yielded an unstable crystal form that gave a rather poor X-ray diffraction pattern. Although it was not possible to derive the complete structure, it was possible to determine the absolute configuration of the molecule at low resolution and without knowing its constitution: a very unusual situation. New attempts to crystallize the caesium as well as the rubidium salt from cold aqueous methanol yielded stable isomorphous monoclinic crystals which gave better X-ray diffraction patterns. From the data collected with the stable monoclinic Rb-salt, the complete structure of the DVB$^-$ (Fig. 1) was derived [2,3]. The absolute and relative configurations of

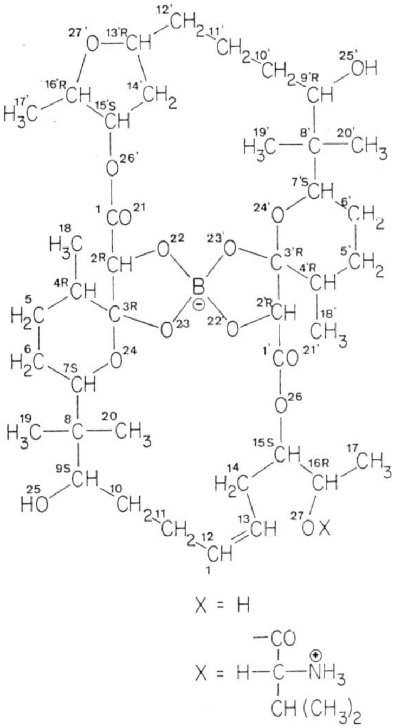

Fig. 1. Constitution and configuration of boromycin and of desvalino-boromycin-anion (DVB$^-$) [2].

asymmetric carbon atoms are specified in this figure by (R) or (S) descriptors. The constitution and configuration, i.e. the primary structure determine the tertiary structure which is represented in Fig. 2. A characteristic feature of this tertiary structure is a cavity, a cleft, in which eight oxygen atoms are arranged in such a way as to complex with a cation of suitable size. The remaining part of the molecule is covered with hydrogen atoms bound to carbon, and is thus lipophilic. The primary structure of DBDVB was deduced from the structure of DVB⁻ and confirmed by X-ray analysis using direct methods [4]. As shown in Fig. 3, the constitution and configuration remain unchanged by acid hydrolysis. However, it is noteworthy also that the tertiary structure shown in Fig. 2 is practically unchanged on removal of the spiro-boron atom of DVB and replacement of the cation by a molecule of water. Only minor changes of torsional angles and mean-square vibrational amplitudes of atomic nuclei were observed. The tertiary structure of DBDVB would appear to make the molecule favourable for esterification by boric acid as well as for complexation with cations of suitable size.

Let us now have a closer look at the constitution and the configuration of DBDVB, which is a macrodiolide, a 32-membered dilactone that can be con-

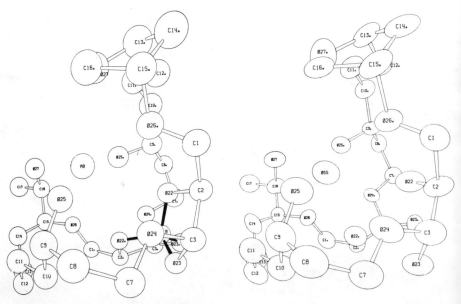

Fig. 2. Perspective view of the main ring in DVBRb and DBDVB (with attached water molecule O(55)). Oxygens are shown but other substituents have been omitted for the sake of clarity [4].

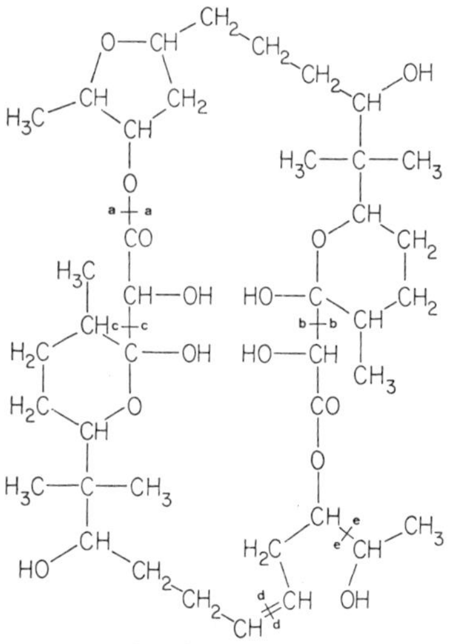

Fig. 3. Constitution of desboro-desvalino-boromycin (DBDVB) [4].

structed from two slightly different, substituted 16-hydroxy-heptadecanoic acids. (Several macrodiolides have been isolated recently from cultures of microorganisms e.g. colletol, colletodiol, colletoketol, pyrenophorol, pyrenophorin, vermiculin and antimycins [8]. These macrodiolids, with 9-, 14- and 16-membered rings, differ in several ways from the 32-membered DBDVB, although their biogenesis may be similar.) One of the hydroxy acids contains a tetrahydrofuran ring, whereas in the other a Δ^{12}-cis-double bond and a hydroxyl at C-16 are present instead. In addition the configurations at C-9 and C-9' are different, whereas all other configurations of corresponding pairs are equal. This is interesting for the following reason: DBDVB contains 15 asymmetric carbon atoms and a cis-double bond, i.e. totally 16 stereogenic units, corresponding to 2^{16} = 65.536 stereoisomers. There is no evidence for constraints which would make any of these stereoisomers incapable of existence, but the one occurring in degradation products of boromycin seems to be particularly favourable for complexing. This is especially remarkable when one considers that any change in the (different!) configurations of C-9 and C-9' would remove oxygens on these atoms from positions predisposed for complexing. The structure of the osmium(VIII)-oxide periodate oxidation products could be easily deduced from the known structure of DBDVB on the basis of their spectral properties. The neutral compound $C_{18}H_{32}O_5$ (Fig. 4a) is a straightforward product of oxidative

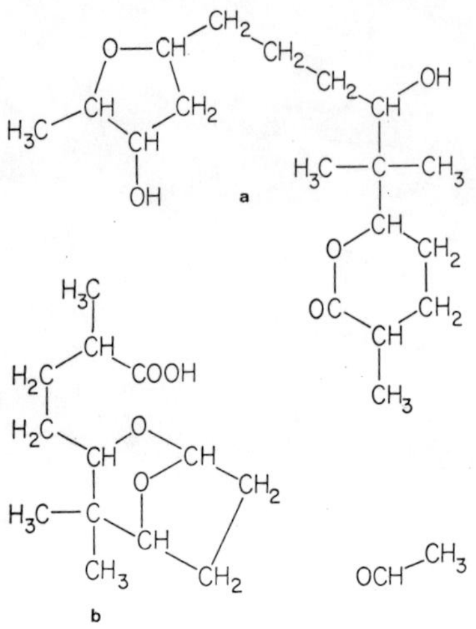

Fig. 4. Products of oxidative degradation of desboro-desvalino-boromycin (DBDVB) [2].

fission at positions a and b. The acid $C_{13}H_{22}O_4$ (Fig. 4b) is produced by oxidative fission at positions c and d and subsequent intramolecular acetalisation of the intermediate aldehyde. The acetaldehyde originates from oxidation at the point e. The structure of boromycin itself was deduced from the structures of its degradation products, assuming that the stable tertiary structure of DVB⁻ is preserved. The D-valine must be bound as ester to C-16, because in DVB⁻ this carbon carries the only hydroxyl which is not in the cavity of the tertiary structure. The ammonium ion of D-valine is presumably complexed by oxygens of the cavity, which is then plugged by the isopropyl group of the amino acid to make boromycin a stable, neutral, highly lipophilic compound. Inspection of the space filling (CPK) models shows not only that the surface of the almost spherical molecule of boromycin is covered with hydrogens bound to carbon, but also that D-valine fits much better than the L-enantiomer.

Boromycin contains two stereogenic units more than DBDVB (the asymmetric carbon atom of D-valine and the asymmetric tetrahedral boron atom of (R)-configuration). Thus $2^{18} = 262,144$ stereoisomers are possible, of which only one has been found so far in Nature.

Structure of aplasmomycin

The structure of another boron-containing natural product, the aplasmomycin $C_{40}H_{60}O_{14}BNa$, has been determined recently by Nakamura et al. [6] by X-ray structural analysis of the corresponding silver salt. The structure shown in Fig. 5 resembles strongly the structure of DVB⁻. The main difference is that the two 'halves' of the aplasmomycin-anion are constitutionally and configurationally identical and correspond to the 'upper half' of DVB⁻. In addition, two trans-double bonds Δ^{11} and $\Delta^{11'}$ are present in aplasmomycin. In contrast to boromycin the configurations at C-9 and C-9' are the same in aplasmomycin; in both molecules the resulting arrangement of oxygens is favourable for complexing. The resemblance of structure matches a resemblance of antibiotic activity: DVB⁻ and aplasmomycin have virtually the same activity against the following tested microorganisms: Bacillus subtilis, Bacillus brevis, Staphylococcus aureus, Clostridium pasteurianum. They are both inactive against Streptomyces viridochromogenes and Mucor [9] (cf. also [7]).

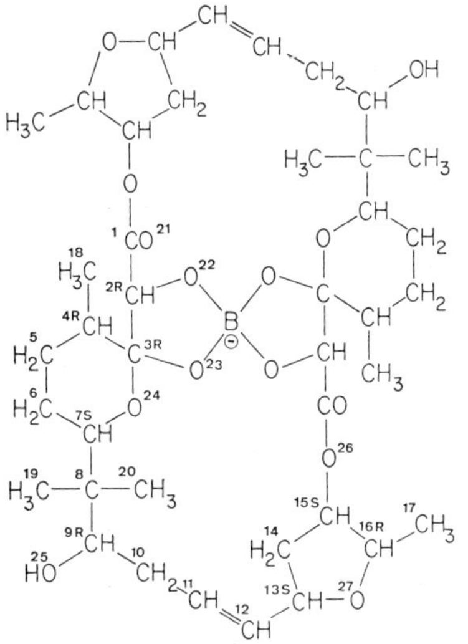

Fig. 5. Constitution and configuration of aplasmomycin [6].

Desvalino-boromycin-anion as ionophore

The structure of the salts of DVB⁻ and their lipophilicity indicate that the anion is an ionophore which transports hydrophilic ions into lipophilic phases. Its hydrophilic cavity, covered with oxygen atoms, complexes readily with cations possessing suitable ionic radii. Hydrolysis of boromycin with tetramethylammonium hydroxide yields a tetramethylammonium salt of DVB⁻. The cation is too large (r = 34.7 pm) to enter the cavity, and hence the salt is not lipophilic but soluble in water and methanol. The conversion of the hydrophilic tetramethylammonium salt with NaCl, KCl, NH_4Cl, RbCl or CsCl gives quantitatively tetramethylammonium chloride and the more stable lipophilic salt-complex of the corresponding cation. Hydrolysis of boromycin with lithium hydroxide gives a salt which is also poorly soluble in lipophilic solvents and soluble in water and methanol. The Li⁺ is too small (r = 6.8 pm) to complex efficiently. Since PMR studies of DVB⁻ salts show that the coupling constants of protons in all these salts are practically the same, the DVB⁻ must be fairly invariant to changing the cation and, hence, fairly rigid. The chemical shifts, on the other hand, depend on the ionic radii of the corresponding cations, as shown in Table 1 for the singlets of protons on C-2 and C-2'. These protons in lipophilic salts of K⁺, Rb⁺ and Cs⁺ are almost isochronous because their local environment in the anion has approximate C_2 symmetry. The Na⁺- and Li⁺-ions evidently lower this symmetry, because they do not fit so well into the rigid cavity. The Me_4N^+ cation which cannot enter the cavity does not lower the symmetry. The chemical shift differences of Na⁺ and K⁺ salts are large enough to be used to observe the exchange rate of these ions. The spectra of an equimolar mixture of both salts in CCl_3D at 30° are superposition of the separate spectra obtained before mixing, which means that the exchange is rather slow.

As already mentioned, space-filling models indicate that D-valine fits better into the cavity of boromycin and its degradation products than its L-enantiomer.

Table 1

r pm	Li⁺ 6,8	Na⁺ 9,8		K⁺ 13,3	Rb⁺ 14,8	Cs⁺ 16,7	Me_4N^+ 34,7
C-2; C-2' τ ppm	4.28 [a] 4.40 [a]	4.42 [a] 4.46 [a]	4.82 [b] 4.88 [b]	4.92 [b] 4.96 [b]	4.95 [b] 4.96 [b]	4.97 [b] 4.98 [b]	4.40 [a] 4.40 [a]
Δτ	0.12	0.04	0.06	0.04	0.01	0.01	0.00

Varian 100 MHz [a] in CD_3OD, [b] in C_6D_6 [10].

This suggested that DVB⁻ might be an enantiomer-selective ionophore. Partition experiments between aqueous and lipophilic phases show that this is indeed so (Fig. 6) [7].

To equal volumes of water and a lipophilic solvent (1,2-dichloroethane was used as standard solvent) known amounts of DVBNMe$_4$ and an excess of racemic salt of the chiral cation are added, and the mixture is then shaken at 4°. The concentration of the cations ($c_R^{aq} + c_S^{aq}$) in the separated aqueous layer is determined from optical density, and the enantiomeric purity is estimated

$$p = \frac{c_R^{aq} - c_S^{aq}}{c_R^{aq} + c_S^{aq}}$$

from circular dichroism. From these data and from the known initial concentrations c_{DVB}^o and $c_R^o = c_S^o$ the ratio

$$\frac{K_{RL}}{K_{SL}}$$

of equilibrium constants of the diastereomeric complexes in lipophilic phase can be calculated. This now characterizes the enantiomer-selectivity of the ionophore in the solvent used at the temperature of the experiment. The results in Fig. 6 show that the DVB is highly selective for D-phenyl-glycine derivatives. However, it does not discriminate between the enantiomers of α-phenylethylammonium-chloride. The ionophoric properties and enantiomer-selectivity of DVB⁻ (and

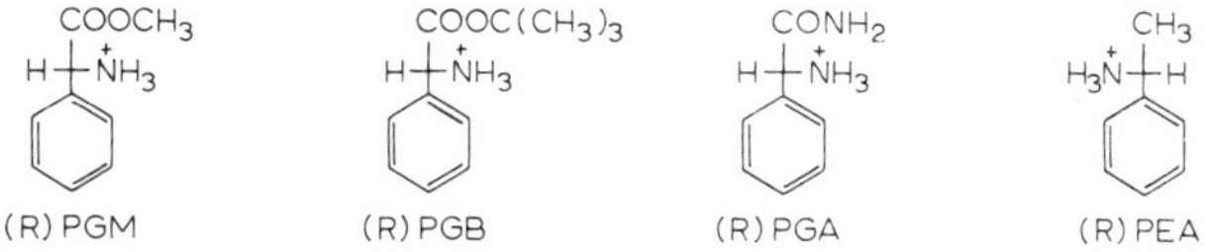

Tetramethylammonium desvalino-boromycin
Solvent ClCH$_2$CH$_2$Cl ; t = 4°C

Cation	K_{LR}/K_{LS}	$\Delta(\Delta G)$ cal
PGM	2.04	-392
PGB	2.27	-451
PGA	2.65	-536
PEA	1.00	0

Fig. 6. Enantiomer-selectivity of desvalino-boromycin-anion (DVB⁻) [11].

presumably of aplasmomycin) are remarkable because this molecule possesses properties encountered usually only with much larger ones, such as enzyme proteins: (a) it has a stable, rigid tertiary structure, determined by the constitution and configuration of numerous chirality centers; (b) the hydrophilic atoms are concentrated in this tertiary structure in a cavity, which is the 'active site' of the ionophore; the surface is lipophilic; and (c) the hydrophilic cavity is enantiomer selective. The question, what is the biological function of such an unusual secondary metabolite, is an intriguing one.

References

1. Hütter R., Keller-Schierlein W., Knüsel F., Prelog V., Rodgers G.C., Jr., Suter P., Vogel, G., Voser W., Zähner H. (1967) Helv. Chim. Acta, *50*, 1533.
2. Dunitz J.D., Hawley D.M., Mikloš D., White D.N.J., Berlin Yu., Marušić R., Prelog V., (1971) Helv. Chim. Acta, *54*, 1709.
3. Nachr. Chem. Techn., *19*, 271 (1971).
4. Marsh W., Dunitz J.D., White D.N.J. (1974) Helv. Chim. Acta, *57*, 10.
5. Okami Y., Okazaki T., Kitahara T., Umezawa H. (1976) J. Antibiot., *29*, 1019.
6. Nakamura H., Iitaka Y., Kitahara T., Okazaki T., Okami Y. (1977) J. Antibiot., *30*, 714.
7. Prelog V. (1978) Plenary Lecture 26th International Congress of Pure and Applied Chemistry, Tokyo, 4–10 September 1977, Pure Appl. Chem., *50*, 893.
8. Keller-Schierlein W. (1978) In: A.I. Laskin, H.A. Lechevalier, Eds.: Handbook of Microbiology, CRC Press Inc., Cleveland, Ohio.
9. Zähner H., Tübingen, private communication.
10. Jurczak J. and Sternhell S., private communication.
11. Žinić M., private communication.

Yu.A. Ovchinnikov and M.N. Kolosov (eds.) Frontiers in Bioorganic Chemistry and Molecular Biology © 1979, Elsevier/North-Holland Biomedical Press

CHAPTER 6

The phalloidin story

THEODOR WIELAND and HEINZ FAULSTICH

Introduction

Phalloidin was the first toxic compound to be isolated in a crystalline state from the poisonous green death cap *Amanita phalloides* in 1937 [1]. When injected intraperitoneally into white mice, 2—2.5 mg of this material per kg bodyweight killed half of the animals. The mice died as soon as 2—4 hours after administration of the toxin. Together with six similar compounds to be discovered in the following years, phalloidin represents a family of quickly acting drugs in the mushroom, the so-called phallotoxins.

Nevertheless, these toxins do not contribute to the fatal intoxications by ingestion of the mushroom. Lethal poisoning by Amanita mushrooms is in all cases solely due to the amatoxins, which represent the second family of toxic peptides in Amanita phalloides. Their lethal dose is about 0.3 mg per kg white mouse and death occurs not earlier than 3—5 days after administration. Obviously, the resorption of amatoxins in the gastrointestinal tract of humans is adequate to produce an intracellular concentration of the toxin, which is sufficient to inhibit the low concentrated (5×10^{-8} M) receptor enzyme, RNA polymerase B (or II) in the liver. This is not the case for phallotoxins; although being preferentially absorbed by the liver cells, oral administration of phallotoxins cannot achieve an intracellular concentration of the toxins high enough to complex totally the cellular actin. Actin of liver cells is the target protein of phallotoxins; in liver cells it has a concentration as high as 5×10^{-5} M, which is a 1000-fold of that of RNA-polymerase B (or II). Consequently, phallotoxins cause lethal lesions only, if high intracellular concentrations are reached, e.g. after parenteral administration.

In spite of its minor part during mushroom poisoning, phalloidin has proven to be an interesting compound from the chemical point of view as well as from its biochemical implications.

97

Structure

The peptidic nature of phalloidin has been recognized already in 1940 [2], but it took almost two decades until a structural formula of the phytotoxin could be established [3,4]. Phalloidin turned out to be a bicyclic heptapeptide; the formula is depicted on top of Table 1. It contains one uncommon amino acid, D-threonine; all the other amino acids are L-configurated; some of them, however, are further derivatized by either hydroxylation (*allo*-L-hydroxyproline, dihydroxy-L-leucine) or oxidative condensation (tryptathionine from tryptophan and cysteine). In Table 1 also the formulae of the six remaining members of the phallotoxin family are listed together with their LD_{50} values. (For A-values see page 106, for references see [5,6].) Obviously all naturally occurring variations in the side chains of the major ring have only little influence on toxicity. Not so has the lack of the hydroxy group of proline-4, which results in a non-toxic peptide, prophalloin [7], a putative precursor of the natural phallotoxins.

Table 1. Naturally occurring phallotoxins, their toxicity (LD_{50} mg/kg for the white mouse, i.p.) and their relative affinity (A) to rabbit muscle actin (phalloidin = 1).

Name	R^1	R^2	R^3	R^4	R^5	LD_{50}	A_{Actin}
Phalloidin	OH	H	CH_3	CH_3	a-OH	2.0	1.00
Phalloin	H	H	CH_3	CH_3	a-OH	1.5	0.74
Prophalloin	H	H	CH_3	CH_3	H	>100	0.01
Phallisin	OH	OH	CH_3	CH_3	a-OH	2.5	0.55
Phallacin	H	H	$CH(CH_3)_2$	CO_2H	a-OH	1.5	0.44
Phallacidin	OH	H	$CH(CH_3)_2$	CO_2H	a-OH	1.5	0.91
Phallisacin	OH	OH	$CH(CH_3)_2$	CO_2H	a-OH	4.5	0.58

Structure and toxicity

Already in 1955, in connection with the sequencing procedure, it had been recognized that monocyclic derivatives of phalloidin are non-toxic. This was

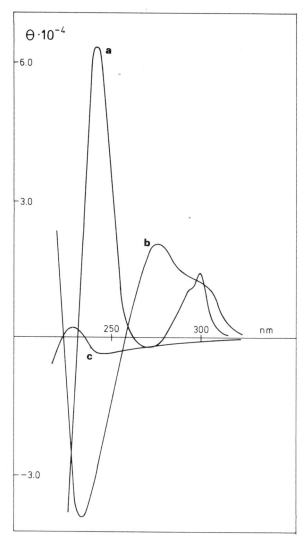

Fig. 1. CD curves in water of phalloidin (a), seco-phalloidin (b) and dethiophalloidin (c).

stated for dethiophalloidin obtained from phalloidin by Raney-Ni-treatment as well as for seco-phalloidin, the monocyclic product of treating phalloidin with acids. Acids will cleave preferably the peptide bond between residues 7 and 1 due to the neighbouring γ-OH group, which easily forms a lactone with the

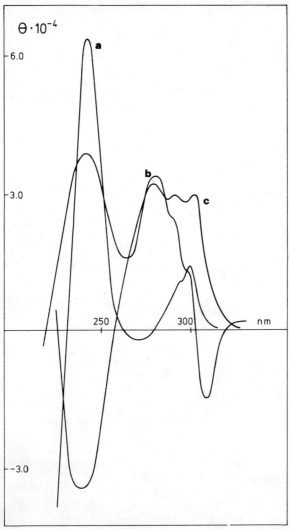

Fig. 2. CD curves in water of phalloidin (a), des-alanine-1-phalloidin (b) and endo-alanine-1a-phalloidin (c).

carboxyl originally involved in the peptide bond. The non-toxicity found for both monocyclic derivatives may be understood on the basis of changes in conformation as documented by the CD-curves (Figs. 1b and c). They differ strongly from that of phalloidin (Fig. 1a) indicating that at least the chromophoric moiety has changed its spatial arrangement.

Early attempts to recognize the contribution to toxicity of the various side chains started with the oxidative degradation of the 1,2-glycol structure in phalloidin by periodate [8]. The product, ketophalloidin, was as toxic as the mother compound and it served as the starting material for many further derivatives. For example, hydrogenation of ketophalloidin by NaBH$_4$ yielded demethylphalloin, which is as toxic as phalloin. Its reaction with 1,2-ethanedithiol gave the toxic dithiolane [9], which after treatment with Raney-Ni led to norphalloin [10]. The latter derivative was even a little more toxic than phalloidin. These derivatives indicate that most of the modifications in the leucine-7 side chain do not affect toxicity.

Only recently, secophalloidin could be subjected to a successful recyclisation [11]. During the cyclisation procedure, the γ-hydroxylated leucine was racemized. Consequently, two products were formed, phalloidin, being identical to

Nontoxic and Toxic Derivatives of Phalloidin

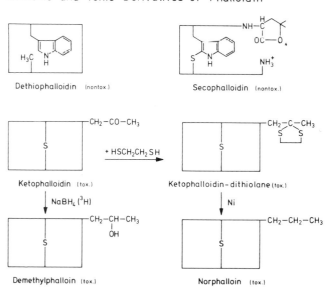

Scheme I. Non-toxic and toxic derivatives of phalloidin.

the natural toxin and an analogue, probably with the epimeric γ,δ-dihydroxy leucine in position 7 [12]. The non-toxicity of the latter proves that a D-configurated amino acid in position 7 deletes the toxicity.

Likewise deleterious is D-alanine in position 1. D-Ala[1]-phalloidin has been obtained by Edman degradation of seco-phalloidin, followed by coupling of D-alanine and recyclisation [13]. By an analogous treatment also the hexapeptide des-alanine-1 phalloidin and the octapeptide endo-alanine-1α phalloidin were obtained, both being non-toxic as a consequence of their changed conformations (see CD-curves in Fig. 2). The successful recyclisation enabled us to prepare further analogues of phalloidin substituted in position 1. Gly[1]-phalloidin is 3–4 times less toxic than phalloidin and Phe[1]-phalloidin even 10 times, while Val[1]-phalloidin is as toxic as phalloidin. These results suggest an optimum length of this side chain of 1–3 carbon atoms.

As the most universal possibility to prepare all kinds of analogues, the total synthesis of the phallotoxins has been elaborated. In 1971, the synthesis of norphalloin was achieved and gave proof to the proposed structure of the phallo-

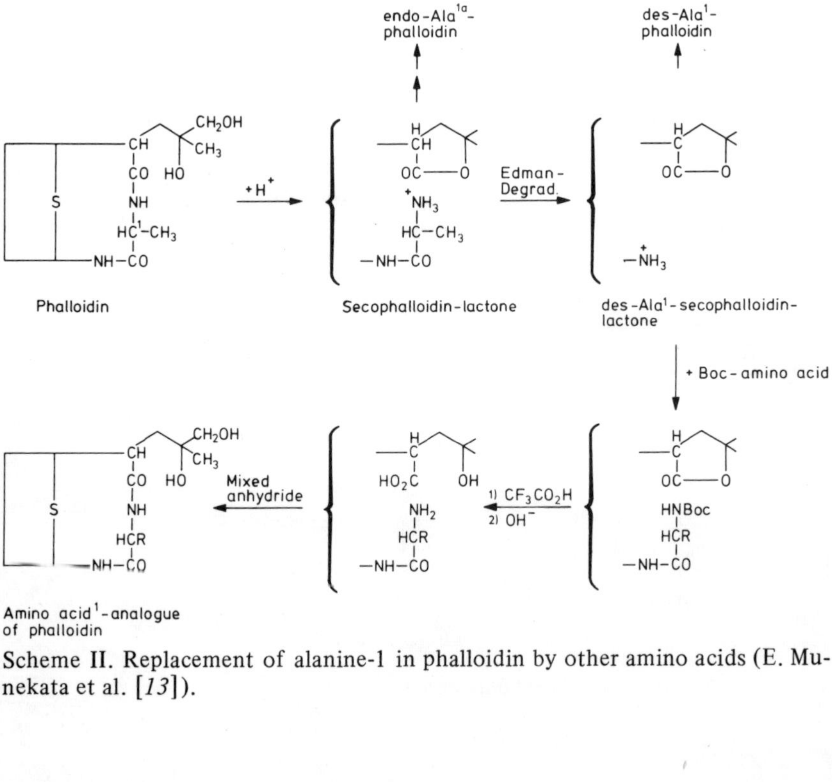

Scheme II. Replacement of alanine-1 in phalloidin by other amino acids (E. Munekata et al. [13]).

Table 2. Toxicity (LD_{50}, mg/kg in white mice) and relative affinity to F-actin of several amino acid-1 analogues of phalloidin.

Amino acid in position 1	Toxicity	Affinity to F-actin (phalloidin = 1,00)
Ala [a]	2.0	1.00
D-Ala	non-toxic [b]	<0.02
Gly	7.5	<0.02
Val	2.5	0.72
Leu	2.5	0.28
Phe	20.0	<0.02
des-Ala	non-toxic [b]	<0.02
Ala-Ala	non-toxic [b]	<0.02

[a] = phalloidin.
[b] = tested in doses up to 30 mg/kg.

toxins [14]. In addition to this, it cleared the way for the synthesis of further analogues. D-Aminobutyric acid-4-norphalloin [15] was toxic and indicated that the hydroxyl group in D-threonine is not essential for toxicity. Glycin-5-norphalloin [16] was completely non-toxic although exhibiting a conformation very similar to phalloidin. This made evident that the methyl group of Ala^5 is involved in binding the toxin to the target protein. Similar conditions apply for the hydroxyl group in *allo*-hydroxyproline, since proline-4-norphalloin is also totally devoid of toxicity. Further synthetic products like hydroxy-proline-4-phalloin and, most recently, proline-4-phalloin [12] confirmed that indeed the hydroxyl group in *cis* position to the carboxyl of proline is involved in the toxin activity. As expected, the leucine-7-analogue [17] of phalloidin was as toxic as the parent compound, indicating once more that the hydroxyl groups in the leucine side chain do not contribute to toxicity. The synthetic phallotoxins are listed in Table 3.

In summary, the *cis*-hydroxyl in *allo*-hydroxyproline-4, the methyl group of alanine-5 and the indolyl moiety of the thioether bridge form the binding site of the toxin to actin, the receptor protein. The participation of the chromophoric indolyl thioether has also been deduced from the difference of optical density in the relevant ultraviolet range, which accompanies binding of the toxin to actin [18]. In agreement with the suggestions mentioned above a molecular model constructed from ^{1}H-NMR data and lowest energy calculations according to Patel et al. [19] shows that the three functional groups mentioned are indeed situated in exposed positions close to each other.

Table 3. Analogues of phalloin obtained by total synthesis.

Analogue	Trivial name	Toxicity (mg/kg, white mouse)	Affinity to F-actin (phalloidin = 1)
OH Leu^7	phalloin (for comparison)	1.5	0.74
OH Leu^{7}Pro^4	prophalloin	>50	<0.02
Leu^7	–	2.0	0.40
Leu^{7}Hyp^4 [12]	–	>100	–
Nva^7	norphalloin	2.0	0.60
Nva^{7}Pro^4	–	>20	–
Nva^{7}D-Abu^2	–	3–4	–
Nva^{7}Gly^5	–	>25	0.02

Nva = norvaline.

Apparently all modifications in the molecule, which move out the indolylthioether from its regular position, will impair the biological activity. Such modifications include opening, contracting or widening of one of the rings. Probably, such a change in the arrangement of the indolylthioether moiety is also induced in one of the oxidation products of phalloidin by peracetic acid [20]. This treatment results in the toxic (R)-sulfoxide and the toxic sulfone, which both exhibit positive Cotton effects in the range of 280 to 300 nm, similar to the parent compound phalloidin. Differently, the non-toxic (S)-sulfoxide has negative Cotton effects in this region. A possible explanation could be that in this sulfoxide the oxygen induces a distortion of the sulfur bridge, which probably changes its inherently helical chirality to that of the antipode. By this the indole nucleus moves to a position, which no longer fits to the binding site of actin.

Interaction of phallotoxins with actin

The interaction of phalloidin with actin was discovered in 1972, when isolated plasma membranes from livers of rats poisoned with phalloidin were examined by electron microscopy [21]. The membrane preparation contained bundles of microfilaments, which were associated with the membrane fragments. Such filaments were only occasionally visible in the membrane preparations of control animals. The filaments behaved unlike actin in that they withstood treatment with chaotropic ions, such as 0.6 M KI. Nevertheless, they were actin, as documented by their reaction with the heavy meromyosin fragment forming the typical arrowhead-like structures [22]. These experiments gave a first hint that phalloidin had interfered with the actin of rat liver cells.

Proof was obtained with rabbit muscle actin, which in presence of phalloidin likewise became stable against chaotropic ions. We learned that such stabilisation represented the specific effect of phalloidin on actin. The fact that liver cell actin and muscle actin behaved very similar with the toxin prompted us to further investigate this interaction with rabbit muscle actin as a model protein. Most of the results obtained with muscle actin have been reviewed in several papers [23] and will be summarized here only briefly.

Phalloidin binds firmly to actin, the K_D of the toxin being as low as 3.6×10^{-8} M [24]. It binds to polymeric or oligomeric actin only, forming a 1 : 1 complex with the protomers in the filaments. No binding has been observed so far with actin monomers. Binding can be followed by difference spectroscopy in the u.v. [18]. The difference spectrum has three maxima at 288, 295 and 305 nm; the latter strongly suggests the participation of the thioether moiety, while the two others may also originate from tyrosine of tryptophan residues in the protein by interaction with the toxin.

In the electron microscope, there is no difference between normal actin filaments and those having complexed phalloidin. However, the adduct phalloidin-F-actin (Ph-actin) differs from F-actin in that it is much more stable against various physical or chemical treatments, which normally depolymerize or destroy filamentous actin.

Among these treatments is 0.6 M KI as already displayed above. Phalloidin even admits a slow polymerisation reaction of G-actin in 0.6 M KI which is not observed in absence of the toxin. Another depolymerizing agent is pancreatic DNAase I. This enzyme forms 1 : 1 adducts with actin monomers, and, as a consequence, shifts the actin equilibrium towards the monomers. There is no depolymerization by this enzyme in the presence of phalloidin, because the interaction of phalloidin with the filamentous form is apparently stronger than that of DNAase I with the monomers.

Phalloidin not only inhibits the depolymerization reaction but also restrains local ruptures in the filaments. Such ruptures can be induced by ultrasonic vibration, by cytochalasin B or by low pH values. The subsequent healing of such perturbations is accompanied by splitting of ATP. Hence the measurement of this weak ATPase activity allows to follow the extent of local ruptures. Phalloidin suppresses such ATPase activities induced by ultrasonication as well as by cytochalasin B completely. Also the ATPase induced by protons is totally inhibited at pH 6, the inhibitory effect decreases with decreasing pH. At pH values lower than 3.5 and in the absence of the toxin the structure of F-actin is so loosened that denaturation occurs, in its presence the structure of the fragments is still stabilized, thus maintaining a relatively high ATPase activity even at low pH values.

Phalloidin protects actin not only from denaturation by protons, but also from denaturation by heat. While F-actin will be denatured by 100% after 3 min heating to 70°C, Ph-actin is denatured by only 15%. Full protection by phalloidin was achieved, when Ph-actin was heated 3 min to 60°C. Similarly, phalloidin protects F-actin filaments from proteolytic digestion as observed with subtilisin, trypsin or α-chymotrypsin. Full protection in this case needs the presence of 0.1 M K^+ ions [25].

As already pointed out, phalloidin shifts the actin equilibrium towards the filamentous form. As a consequence, the concentration of actin monomers in equilibrium with actin polymers decreases. The concentration of monomers, in equilibrium with filaments, $[G]_c$ is, in 0.1 M KCl, about 10^{-6} M. In the presence of 1 equivalent of phalloidin this value decreases to 3.6×10^{-8} M, which is about 30 times lower. In presence of two equivalents of phalloidin the concentration of the monomers further decreases to a value even 100 times lower than without phalloidin. Most probably, the low concentration of monomers can account for most of the stabilization effects observed with muscle actin. For example, proteolytic digest may predominantly affect monomers or terminal protomers in the filaments, which can be expected to be better substrates for proteases. Also heat denaturation probably proceeds via actin monomers, since at room temperature the monomers are less stable than the actin filaments. In these cases the diminution of the quantity of monomers or the decrease of breaks in the filaments as affected by phalloidin could explain the protective capacity of the toxin.

The protective capacity of all phallotoxins investigated so far runs grossly parallel to their affinity towards actin. Such affinity values, related on phalloidin ≡ 1.0, are listed in the Tables 2 and 3. Obviously, the highest affinity to actin is achieved by phalloidin and phallacidin, indicating that the two hydroxyl groups at the leucine side chain play an — albeit minor — role in binding to the protein. In all cases, the affinity to actin of phallotoxins could be correlated with their toxicity in the white mouse. Derivatives, with dissociation constants up to ten times higher than that of phalloidin, were still toxic, while those with dissociation constants between ten and hundred times that of phalloidin turned out to be non-toxic. It is, therefore, reasonable to assume that the binding of the phallotoxins to cell actin, by the concomitant decrease of the concentration of monomeric actin, represents a possible mechanism for phalloidin toxicity.

Toxicity

The toxicity of phalloidin [6] has been under investigation since 1938, when the crystalline toxin became available. It has fascinated both pharmacologists and biochemists up till now. Most of the studies have been done in the rat. Here, phalloidin causes a severe swelling of the liver; in the end the weight of the liver is 2–3 times that of a normal rat. Such swelling is due to an excessive accumulation of blood in this organ, the blood content being five times higher than that in controls. The following depletion of fluid in the periphery probably leads to death of the animals in a hemorrhagic shock.

Such toxic lesions were observed exclusively in the liver. This is a consequence of the specific accumulation of the toxin in this organ [26]. In less than ten minutes, the rat liver takes up more than 70% of the toxin present in the circulatory system. The mechanism of such rapid and specific uptake is still unknown. For comparison, the rate of uptake for amatoxins is about 20 times slower. In addition to this, phallotoxins cannot be excreted into the bile like the amatoxins, because the bile flow stops a few minutes after administration of phalloidin.

In search of the cellular events preceding the accumulation of blood in the liver, a study by electron microscopy established that the hepatocytes become interspersed with vacuoles. These vacuoles develop from invaginations of the membranes near the sinusoidal space, as indicated by erythrocytes and other blood components present in the interior of the vacuoles. An event very similar to this occurs, when isolated hepatocytes are incubated with phalloidin. In this case exvaginations develop, instead of invaginations giving rise to numerous bulbous protrusions or blebs on the cell surface [27] (Fig. 3). In both cases the plasma membrane of the hepatocytes may be regarded as to have lost its inherent elasticity. As a consequence, the membrane can be deformed by even low pressure gradients such as the moderate pressure of the portal veins in vivo or a slight intracellular pressure apparently present in the isolated hepatocytes.

In the perfused rat liver, the vacuolisation induced by phalloidin is succeeded by a series of other damages. One of the earliest events is a loss of K^+ ions from the hepatocytes, which is, later on, followed by a loss of cytoplasmic enzymes. About one hour later the liver is obviously depleted from ATP and glycogen, damages which finally destroy the organ. Anyhow, all the disturbances cited here seem to be consequences of the observed vacuolisation rather than its origin.

In order not to misinterpret 'the injuries induced by phalloidin in hepatocytes, it should be noted here that neither the blebs on the isolated cells, nor the vacuoles in the perfused liver represent lethal damages. For example, hepatocytes covered with blebs have intact membranes as indicated by the exclusion of trypan blue; furthermore, they exhibit no or only minor differences from normal

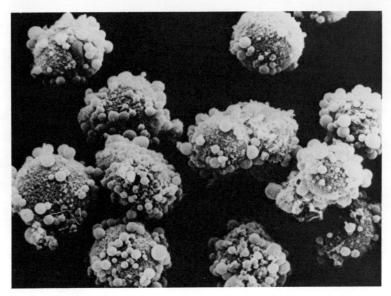

Fig. 3. Isolated rat hepatocytes after treatment with phalloidin. (By courtesy of Dr. M. Frimmer.)

cells with respect to oxygen consumption or glucagon response. Concerning these protrusions on isolated hepatocytes, morphologists [28] have described such bulbous extensions as 'zeiotic blebs', e.g. on K.B. cells. According to these workers, such blebs represent only one variety of surface expression in cells, among others, like microvilli, ruffles and filipodia.

Similarly, also the vacuolisation in the perfused rat liver is not a phalloidin-specific phenomenon. It could be induced also in unpoisoned livers by high perfusion pressures. However, both structural deformations, the zeiotic blebs as well as the pressure-induced vacuoles in the liver, can be reversed by the cells: zeiotic blebs, which need only eight seconds to develop, can be withdrawn by the cell in about one minute. Similarly, the pressure-induced vacuoles can be removed by the hepatocytes in about 30 minutes. Not so the analogous changes induced by phalloidin, which persist for hours. Therefore, it is reasonable to suggest that the toxic activity of phalloidin is represented by its ability to deprive hepatocytes from their capacity to remove such membrane deformations like blebs or vacuoles.

Possible molecular mechanisms of toxicity

Certainly the dynamics of membrane structures involve contractile proteins like actin and it is tempting to apply those observations as done with phalloidin and muscle actin, on non-muscle cells like hepatocytes. Actin in those cells is located in the so-called microfilamentous web, which underlies the plasma membrane. Probably, together with myosin, these microfilaments form a system which can produce mechanical forces. From experiments with muscle actin it is known that the interaction of actin and myosin is not disturbed by phalloidin; for instance, myosin fragments bind to actin filaments although complexed with phalloidin, and Ph-actin activates the myosin-ATPase as efficiently as normal F-actin. Therefore, the disturbance by the toxin will not affect the process which produces mechanical forces by hydrolysis of ATP. Rather, phalloidin disturbs the organisation of such a mechanical system at places inside the cell, where mechanical force is needed.

It is reasonable to assume that hepatocytes use their actin which is present in 5×10^{-5} M concentration, at many places inside the cells by a steady reorganisation of microfilaments. It may well be that such reorganisation represents even a good deal of the dynamics of a cell. Reorganisation of microfilaments, however, needs the translocation of actin. Such a translocation of actin rationally takes place by a depolymerisation of F-actin, followed by a repolymerisation at places where the cell requires it. It is evident that such a process depends on the concentration of actin monomers, and that it would be badly disturbed by phalloidin which can decrease the concentration of actin monomers by a factor of 100. We estimate that the translocation of actin inside the cell which normally takes a few minutes, will last for hours if phalloidin is present.

Some experimental results collected recently support that theory. One of these results is that blebs on isolated hepatocytes develop only in the case that an excess of phalloidin over cell actin is incorporated into the cells. Provided that most of the actin is in the microfilaments, no toxic activity is to be expected as long as the phalloidin concentration does not exceed the concentration of actin filaments, because their mechanical function seems not to be inhibited by the toxin. Symptoms of toxicity should occur only if the phalloidin concentration is high enough to cause monomeric actin in the cell to polymerize. If most of the actin in the cell is in the polymerized state, this would mean that almost one equivalent of toxin could be incorporated into the cell without any effects. First effects would occur not before nearly all actin is complexed. It is in agreement with this theory that also excess of toxin can enlarge the toxic effects simply by mass action. This has been observed in the above experiment.

A second experiment gives almost a direct evidence of phalloidin inhibiting

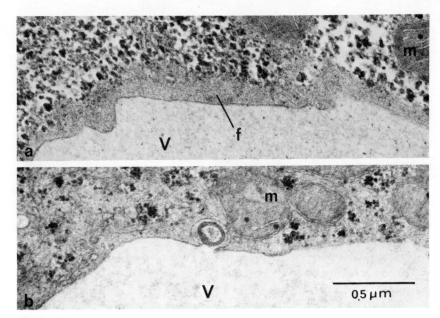

Fig. 4 (a and b). Endocytotic vacuoles (V) in parenchymal cells of a perfused rat liver. Vacuoles induced by posthepatic pressure are surrounded by a filamentous web (f). This filamentous web is absent on vacuoles induced by phalloidin (Fig. 4b) (m − mitochondrion, 50.000×). (By courtesy of Dr. W. Jahn.)

the translocation process of microfilaments in hepatocytes [29]. Such proof was obtained when rat livers with a pressure-induced vacuolisation were compared to others poisoned with phalloidin. The pressure-induced vacuoles are seamed with a layer of microfilamentous material which most obviously is actively involved by extensive formation of microvilli, when the vacuoles disappear. Different from this, the vacuoles induced by phalloidin lack the surrounding microfilamentous web and hence persist (Fig. 4). Evidently these cells are not able to translocate microfilaments from other places, e.g. the surroundings of bile canaliculi, where is abundance of them, to the vacuoles, where it is urgently needed to remove the membrane deformation.

Besides the possibility explained in details above, there may exist still other possibilities for the molecular mechanism by which phalloidin disturbs structures or dynamics of a liver cell. Phalloidin evidently does not impair the interaction of actin with myosin. However, there is only little experience of what other proteins which normally bind to actin like α-actinin or tropomyosin, do in presence of phalloidin. Changes in their affinity to actin could likewise account for the observed deformations in the cell membranes. A further possibility is that actin, besides its mechanical functions in the cell, may exert also regulatory functions still unknown, which could likewise become affected by phallotoxins. However, among these theories, the possibility that phalloidin simply acts by a fatal decrease of actin monomers, is to date the most promising one.

References

1. Lynen F., Wieland U. (1937) Justus Liebigs Ann. Chem., *533*, 93.
2. Wieland H., Witkop B. (1940) Justus Liebigs Ann. Chem., *543*, 171.
3. Wieland Th., Schön W. (1955) Justus Liebigs Ann. Chem., *593*, 157.
4. Wieland Th., Schnabel H.W. (1962) Justus Liebigs Ann. Chem., *657*, 225.
5. Wieland Th. (1968) Science, *159*, 946.
6. Wieland Th., Wieland O. (1972) The toxic peptides of Amanita species. In: S. Kadis, A. Ciegler, S.J. Ajl, eds.: *Microbial Toxins*, Vol. 8, Academic Press, New York, N.Y.
7. Munekata E., Faulstich H., Wieland Th. (1978) Justus Liebigs Ann. Chem., 776.
8. Wieland Th., Schöpf A. (1959) Justus Liebigs Ann. Chem., *626*, 174.
9. Wieland Th., Rehbinder D. (1963) Justus Liebigs Ann. Chem., *670*, 149.
10. Wieland Th., Jeck R. (1968) Justus Liebigs Ann. Chem., *713*, 196–200.
11. Munekata E., Faulstich H., Wieland Th. (1977) Angew. Chem., *89*, 274–275; (1977) Angew. Chem. Int. Ed. Engl. *16*, 267–268.
12. Munekata E., personal communication.
13. Munekata E., Faulstich H., Wieland Th. (1977) J. Amer. Chem. Soc., *99*, 6151–6153.
14. Fahrenholz F., Faulstich H., Wieland Th. (1971) Justus Liebigs Ann. Chem., *743*, 83.
15. Heber H., Faulstich H., Wieland Th. (1974) Int. J. Pept. Protein Res., *6*, 38.
16. Faulstich H., Nebelin E., Wieland Th. (1973) Justus Liebigs Ann. Chem., *1973*, 50.
17. Munekata E., Faulstich H., Wieland Th. (1977) Justus Liebigs Ann. Chem., 1758.
18. Wieland Th., de Vries J.X., Schäfer A., Faulstich H. (1975) FEBS Letters, *54*, 73.
19. Patel D.J., Tonelli A.E., Pfaender P., Faulstich H., Wieland Th. (1973) J. Mol. Biol., *79*, 185.
20. Faulstich H., Wieland Th., Jochum Chr. (1968) Justus Liebigs Ann. Chem., *713*, 186–195.
21. Govindan V.M., Faulstich H., Wieland Th., Agostini B., Hasselbach W. (1972) Naturwissenschaften, *59*, 521.
22. Lengsfeld A.M., Löw I., Wieland Th., Dancker P., Hasselbach W. (1974) Proc. Natl. Acad. Sci. USA, *71*, 2803.
23. Wieland Th. (1977) Naturwissenschaften, *64*, 303–309.

24. Faulstich H., Schäfer A.J., Weckauf M. (1977) Hoppe-Seyler's Z. physiol. Chem., *358*, 181—184.
25. DeVries J., Wieland Th. (1977) Biochemistry, *17*, 1965.
26. Faulstich H., Wieland Th., Schimassek H., Walli A.J., Ehler N., Mechanism of phalloidin intoxication in membrane alterations as basis of liver injury (1977) In: Popper H., Bianci L., Reuter W., eds.: *Falk-Symposium 22*. MTP Press, Lancaster, p. 301.
27. Weiss E., Sterz I., Frimmer M., Kroker R. (1973) Beitr. Pathol., *150*, 345.
28. Kessel R.G., Shih C.Y. (1974) *Scanning Electron Microscopy in Biology*. Springer, Berlin, Heidelberg, New York, p. 71.
29. Jahn W. (1977) Cytobiologie, *15*, 452.

CHAPTER 7

Antiviral, antimitogenic and antimalarial activities of synthetic analogues of S-adenosyl-homocysteine

MALKA ROBERT-GÉRO and EDGAR LEDERER

Introduction

Biological transmethylation reactions, discovered by Du Vigneaud in 1939 [1], are catalysed by enzymes, called methylases (or transmethylases), which transfer the methyl group of S-adenosyl-methionine, discovered by Cantoni in 1952 [2], to nucleophiles such as –OH, –NH, –SH groups, or C=C double bonds. (For a review on biological C-alkylation reactions see [3]).

In recent years molecular biologists have discovered a large number of enzymatic methylation reactions of macromolecules; especially nucleic acids [4,5] and proteins [6,7]. These lead to methylated purine and pyrimidine bases in all categories of nucleic acids, to 2'-O-methylated nucleosides in some types of RNA [8], and to N- or carboxymethylated aminoacids in proteins; these methylation reactions are common to all living organisms and are of fundamental importance for cell division, differentiation and the normal functioning of the cell.

Recent investigations have shown that all tRNA molecules [9–11] and most mRNA molecules [12–14] have a specific pattern of methylated bases, which is necessary for full biological activity; the same is true for ribosomal RNA [15] and for eukaryotic DNA [16–18]. DNA methylases in prokaryotes have the important function of methylating a few specific bases, thus protecting the molecule from hydrolysis by the so-called restriction nucleases [19–21].

The field of protein methylation (so well developed and summarized by Paik and Kim [6–7]) has recently gained still more interest by reports from Adler's [22,23] and Koshland's [24,25] laboratories showing that bacterial chemotaxis is dependent on the specific enzymatic methylation of the γ-carboxyl groups of some glutamic acid residues of a surface protein of E. coli.

Inhibitors of biological methylation reactions

Since the discovery of abnormal methylation patterns of nucleic acids in tumors [26–28] it became obvious that inhibitors of transmethylases might have interesting effects. S-adenosyl-homocysteine (SAH), one of the products of S-adenosyl-methionine (SAM) dependent transmethylases is a strong inhibitor of nearly all these enzymes [29–30].

Other *natural inhibitors* of transmethylases have also been described, such as nicotinamide [31] or glycine [32,33], both of which seem to act by accepting methyl groups from SAM, thus diminishing its concentration and at the same time generating SAH. Cytokinins, such as kinetin riboside, or zeatin riboside have also been reported to inhibit tRNA methylases [34].

More recently, Barbier et al. [35] have shown that various tRNA methylases are inhibited by purified extracts from the androgenic glands of the crab (Carcinus maenas) and even by t,t-farnesylacetone [36], one of the constituents of the purified extracts; it is still so be seen, whether the androgenic hormone inhibits protein synthesis in the ovaries of the crab via inhibition of tRNA methylation.

Amongst several more or less well-defined macromolecular inhibitors of transmethylation described in the literature, *interferon*, an antiviral glycoprotein is undoubtedly the most interesting. Lebleu et al. [37,38] have shown that it inhibits directly or indirectly the methylation of viral messenger RNA. As we shall see, the synthetic viral inhibitors we have studied may act by an analogous mechanism.

Synthetic inhibitors of transmethylases have been studied by various laboratories. Such inhibitors can be either:
— *analogs of the substrate*: inhibition of catechol-*O*-methyltransferases by sulfur analogs of dopamine and norepinephrine [39] or by hydroxylated naphtoquinones [40,41], inhibition of norepinephrine *N*-methyltransferase by benzylamines [42] or benzamidines [43], of adrenal phenethanolamine-*N*-methyl transferase by substituted imidazols [44], or of the biosynthesis of ergosterol by 25-aza-zymosterol [45], etc., or
— *analogs of SAH*: several groups have prepared synthetic analogs of SAH and have tested them on purified methyltransferases in vitro [46–49,50–53,54, see also 55, 56; for a review see 57].

In our laboratory, following a suggestion of Dr. Jean Hildesheim, a series of simple, structural analogs of SAH was prepared and their inhibitory activity in vitro on rabbit liver methylases studied [58–63]. Some of these, such as

5'-deoxy-5'-S-isobutyl-adenosine (SIBA) were inhibitory, but SAH was at least 10 times more active than the synthetic analogs, confirming in general the results of other authors that, in vitro, purified methylases are most strongly inhibited by SAH, the natural inhibitor [57].

Thus, it was all the more astonishing and satisfying, when one of us (M. Robert-Géro) found that these same compounds, especially SIBA, which was only slightly toxic for normal chick embryo fibroblast in culture, inhibited very strongly the oncogenic transformation of these cells by Rous sarcoma virus (RSV) and virus multiplication in cell culture. SAH was inactive under the same conditions [64,65].

The inhibitory activity of SIBA in cell culture was then confirmed with other oncogenic RNA and DNA viruses [66,67] and led to a more detailed study of the biological activities of SIBA and its analogues.

In the following we shall give some details of the various antiviral activities of SIBA and its analogues and shall try to define their mechanism of action. An obvious possibility is that they inhibit methylation of viral mRNA and thus stop viral multiplication and cell transformation.

Having had such an unsuspected success with antiviral activities of SIBA, we

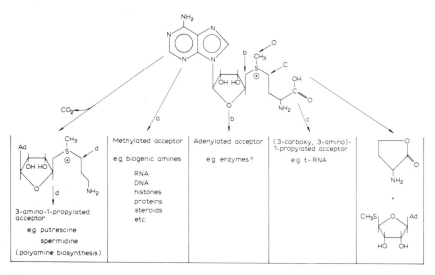

Fig. 1. Possible modes of nucleophilic attack on S-adenosyl-L-methionine. (J.K. Coward, in: *The Biochemistry of Adenosylmethionine*, F. Salvatore, E. Borek, V. Zappia, H.G. Williams-Ashman and F. Schlenk, eds., Columbia University Press, New York, N.Y., 1977, p. 127.)

Table 1: Biological activity of synthetic SAH analogues.

No.	R =	Abbrev.	Conc. mM	% Inhibition of cell transformation	Toxicity to normal cells*	Ki tRNA methylase µM	Ki protein methylase I µM	Ki DNA methylase µM
1	HOOC−CH(NH$_2$)(CH$_2$)$_2$−S−	SAH-L	1.0	19.5	−	1.5	1.5	0.23
		SAH-D	1.0	56	−	−	−	−
2	CH$_3$S−		0.5	95	+	13600		
3	CH$_3$(CH$_2$)$_2$S−		0.5	83	±	8		
4	CH$_3$(CH$_2$)$_3$S−		0.5	86	+			
5	(CH$_3$)$_2$CH−CH$_2$−S−	SIBA	0.5	100	−	3600	635	1420
6	(CH$_3$)$_2$CH−CH$_2$−S→O		1.0	75	±			
7	(CH$_3$)$_2$CH−CH$_2$−S− N-6 benzoyl		0.5	99	−			
8	CH$_3$−CH$_2$−CH−S− | CH$_3$	Iso-SIBA	0.5	99	−	1060	376	
9	CH$_2$=CH−CH$_2$−S−		0.5	99	±	8		
10	CH$_2$=C−CH$_2$−S− | CH$_3$		0.25	97	−	370		
11	CH$_3$(CH$_2$)$_6$−S−		0.5	toxic	++	8		8
12	HOCH$_2$CH$_2$−S−		1.0	96	−	5100		375
13	NH$_2$(CH$_2$)$_2$−S−		1.0	87	−			
14	HOOC−CH(NH$_2$)CH$_2$−S−		1.0	76	±	680		174
15	⟨N⟩−CH$_2$−S−		0.5	toxic	++	10000		
16	CH$_3$−S−CH$_2$−S−		0.5	100	−	6800	320	

*) Cell viability after treatment: − = 80-100 %; ± = 60-80 %; + = less than 50 %; ++ = 0

Table 1 and 2: *Effect on normal cell multiplication.* Secondary cultures of chick embryo fibroblasts were seeded at 5 × 10^5 cells/dish and 24 hours later the analogues were added at the desired concentration. The cultures (in duplicate) were incubated at 37°. The cells from the control cultures and from the treated cultures were counted every 24 hours.

To distinguish between cytostatic and cytotoxic effect, the inhibitor-containing media were replaced by the standard growth medium after various incuba-

Table 2: Biological activity of synthetic SAH analogues.

No.	R =	Conc mM	% Inhibition of cell transformation	Toxicity to normal cells
17	HOOC-CH(NH$_2$)(CH$_2$)$_2$S- N-6 methyl	1.0	22	-
18	(CH$_3$)$_2$CH-CH$_2$S- N-6 methyl	0.5	86	-
19	CH$_3$CH$_2$-CH(CH$_3$)-S- N-6 methyl	0.5	80	-
20	(CH$_3$)$_2$CH-CH$_2$-S(O)$_2$-	1.0	0	-
21	(CH$_3$)$_2$CH-CH$_2$-S$^+$(CH$_3$)-	1.0	19	-
22	CH$_3$C(O)-NH(CH$_2$)$_2$S-	1.0	19	-
23	(CH$_3$)$_3$N$^+$-(CH$_2$)$_2$S-	1.0	0	-
24	HOCH$_2$CH(OH)CH$_2$-S-	0.5	70	-
25	CH$_3$CH(OH)CH$_2$-S-	0.5	87	-
26	HOOC-CH$_2$CH$_2$-S-	1.0	59	±
27	(CH$_3$)$_2$CH-CH$_2$-S- 2' and 3' phosphite	0.5	82	-
28	CH$_3$(CH$_2$)$_6$-S-	0.5	42	++
29	CH$_3$(CH$_2$)$_6$-S(O)-	0.5	54	-
30	CH$_3$(CH$_2$)$_9$-NH(CH$_2$)$_2$-S-	1.0	0	-

tion times, and cells were counted again one and two days later. A product was considered as only cytostatic if, upon renewal of the medium 24 or 48 hours later, the cell number doubled.

Inhibition of cell transformation. Secondary CEF cultures were plated as before, and infected one day later with 100 FFU of RSV. Cultures were then overlaid with 0.8% Difco agar. Cytostatic concentrations (for normal cells) of the inhibitors in 1.5 ml of liquid medium were added on top of the gelled underlayer, immediately after virus adsorption. Control cultures were overlaid with inhibitor-free medium. After two days of exposure, the inhibitor-containing liquid overlayer was replaced by the standard growth medium. Foci of transformed cells were counted eight days later and their number compared with the number of foci from control cultures.

The preparation of the cell-free extracts and the determination of the tRNA methylase activity was the same as described in a previous publication [64]. The protein methylase activities were measured as indicated by Paik and Kim [7]. The concentration of SAM was always 10^{-4} M in the assays. DNA methylase activities were measured in rat hepatocytes (A. Berneman, unpublished).

For the preparation of the analogues see [58,59,63,70].

thought it might also act as an antimitogen in mitogen-stimulated lymphocytes. This was found to be true [68].

And why not try to stop an 'infective messenger' in other situations, for instance in the case of intracellular parasites? This is indeed possible, as Dr. W. Trager of the Rockefeller University, New York, has shown that SIBA inhibits the multiplication of the malaria parasite, Plasmodium falciparum, in human erythrocytes [69].

We are, of course, aware of the possibility that SIBA and its derivative are not only SAH but also SAM analogues, and a glance at Fig. 1 which describes most of the SAM-dependent reactions, shows the complexity of the problem, which might also be related to polyamine metabolism, to cell permeability, etc. . .

Screening of antitransforming activity in cell culture

Forty synthetic analogues of SAH have been tested on chicken cell fibroblasts (CEF) in culture, infected with Rous sarcoma virus. Tables 1 and 2 show the percent inhibition of cell transformation obtained with 30 of these (including SAH for comparison).

Surprisingly, SAH, the natural inhibitor of transmethylases has no effect on focus formation (Table 3). The reason of this ineffectiveness is its rapid intracellular degradation [65]. D-SAH gives 65% inhibition.

For the most active compounds K_i values for tRNA methylase, protein methylase and DNA methylase are also indicated in Table 1.

As shown in Table 1, several compounds strongly inhibit cell transformation but some of them have cytotoxic effects on normal cells. The best results are obtained with compounds having a short 3- to 4-carbon side chain. The introduction of a second sulfur atom in this side chain, such as in compound **16**, maintains or even increases the oncostatic activity.

The N-6 methyl derivatives **17**, **18** and **19** of SAM, SIBA and 5'-deoxy-5'-S-(2-methylpropyl)-adenosine (ISO-SIBA) were prepared, in the hope that they would not be substrates for a deaminase and thus be more stable in the cell, but they were found to be less active than the parent compounds. N-6 methyl SAH was nearly inactive; on the contrary, N-6 benzoyl SIBA (**7**) was nearly as active as SIBA.

The sulfone of SIBA (**20**) was entirely inactive: the S-methyl sulfonium derivative of SIBA (**21**) and N-acetyl-cysteaminyl-adenosine (**22**) were nearly inactive (Table 2).

Compounds with hydrophilic groups such as **23, 24, 25, 26, 27**, or with lipophilic groups, such as **28, 29, 30**, were less active or inactive (Table 2).

Table 3. Effect of SIBA and SAH on focus formation by RSV [64].

	Number of foci [a]	% Inhibition	Number of foci [a]	% Inhibition	Number of foci [a]	% Inhibition
Control	72		110		72	
SAH 1 mM	58	19.5	112	0	60	17
SIBA 0.5 mM	23	68	60	45	2	97
SIBA 1.0 mM	0	100	6	95	0	100
		A		B		C

[a] After 48 hrs of inhibition medium was changed and foci were counted 8 days later.
A = inhibitor added 1 hr after infection,
B = inhibitor added 2 days after infection,
C = inhibitor added 4 days after infection.

Only adenine-containing compounds have been found active: S-isobutyl-inosine being very weakly inhibitory, and S-isobutyl-uracil and S-isobutyl-cytidine inactive.

As 5′-deoxy-5′-S-isobutyl-adenosine (5) was one of the most potent and least toxic analogues, and easy to prepare, its biological activity and mechanism of action were examined in detail.

More recently iso-SIBA (8) and the 2-pyridyl-analogue (15) also have been studied.

Activity of SIBA and iso-SIBA on oncogenic viruses

Table 4 gives a general view of the biological activities of SIBA studied so far, and shows that it is not only active against RSV in chick embryo fibroblast, but also against other RNA viruses, such as mouse sarcoma virus (MSV) (Table 5) and mouse mammary tumor virus (MMTV).

After preliminary experiments in vitro, in cell culture, Dr. J.C. Chermann (Institut Pasteur) has also tested SIBA and iso-SIBA in mice in vivo against Friend's leukemia virus. As seen in Table 4, one dose of 1 mg SIBA prolongs survival time by about 42%. In these experiments, iso-SIBA was found to be somewhat more active than SIBA.

Oncogenic DNA viruses are also susceptible to the action of SIBA; 50 μM SIBA inhibits polyoma virus replication in mouse embryo cells to 98% (Table 4) [66]. Herpes simplex virus in culture of a human epidermoid carcinoma is also inhibited to 98%, but this activity is reversible [67].

Table 4. Biological activities of SIBA (S-isobutyl-adenosine) on cell cultures in vitro and on mice in vivo.

	Active dose	%Inhibition	References
I) *Antiviral activities*			
a) RNA viruses			
1) Rcus sarcoma virus (RSV) transformation of chicken fibroblasts in vitro	0.5 mM (500 μM) 1.0 mM (1000 μM)	70% 100%	[64]
2) Mouse sarcoma virus (MSV)	2 mM 20 μM	~80% 50%	Yoshikura, Orsay, and J.C. Chermann, Inst. Pasteur, unpublished
3) Mouse mammary tumor virus (MMTV) survival of transformed cells	5 μg/ml 10 μg/ml	~50% ~90%	M. Crépin, Inst. Pasteur, unpublished
4) Friend's virus in vivo in mice	1 mg per mouse	42.5% prolongation of survival	J.C. Chermann, Institut Pasteur, unpublished
b) DNA viruses			
5) Polyoma virus mouse embryo cells plaque formation	10 μM 50 μM	92% 98%	[66]
6) Herpes virus human epidermoid carcinoma no. 7 (Hep 2) in cultures infected by herpes simplex virus; virus replication	1 mM	98% after 24 h but reversible	[67]
II) *Antibacterial activity*			
Growth of M. smegmatis on Dubos medium	0.5 mM 1.0 mM	24 h retardation 8 days retardation	G. Farrugia, ICSN, Gif sur Yvette, unpublished
III) *Antimitotic activity*			
Inhibition of blastogenesis of lymphocytes stimulated by mitogens (Con A)	100 μM 300 μM	64% 94%	[68]
IV) *Antiparasitic activity*			
Plasmodium falciparum (malaria)	100–300 μM	50–90%	[69]

Table 5. Oncogenic transformation of mouse cells by a murine sarcoma virus.

SIBA	Time of contact	No. of transformed foci [a]
0	24 hours	Confluent foci
2 mM	24 hours	148
2 mM	24 hours	139
0	48 hours	Confluent foci
0.5 mM	48 hours	218
0.5 mM	48 hours	188
2 mM	48 hours	18
2 mM	48 hours	35

[a] Counted three days after elimination of the inhibitor from the cultures. (H. Yoshikura, Fondation Curie, Institute du Radium, Orsay, unpublished.)

No antiviral activity of SIBA in mM concentration was found with polio virus (Horodniceanu, personal communication) nor with L1210 leukemia cells (Chermann, personal communication).

Antimitogenic activity of SIBA

Lymphocytes in culture can be induced to multiply by the action of lectins (PHA, ConA, etc.). This mitogenic activity can be easily measured in culture by the incorporation of tritiated thymidine. Following our general hypothesis that most phenomena of cell division and cell differentiation depend on several essential (and often not yet recognized) methylation steps, we suggested to Dr. C. Bona (Institut Pasteur) to try the action of SIBA on mitogen-induced blastogenesis of lymphocytes. As expected, he could show that 100–300 μM SIBA prevents lymphocyte multiplication. Addition of 300 μM SIBA, even 1–2 days after stimulation of the cells by ConA or a water-soluble mitogen from Nocardia, strongly inhibited the incorporation of ^{3}H thymidine. After elimination of the inhibitor from the cultures [68], lymphocytes recovered their ability to be stimulated by T or B mitogens.

Antimalarial activity of SIBA

The same 'general hypothesis' as mentioned above led us also to suggest to Dr. W. Trager (Rockefeller University) to test SIBA on his Plasmodium falciparum cultures growing in human erythrocytes: here again, as expected, SIBA strongly inhibits the multiplication of the parasite (Table 6) [*69*].

We do hope that these encouraging results will be extended to other parasitic infections and will lead to in vivo experiments.

Mechanism of action of SIBA and analogues

The experiments in cell cultures described above, are in apparent contradiction with in vitro studies made with SAH analogues using purified methylases; in all these experiments (reviewed by Borchardt [*57*]) SAH was the most active compound and only slight alterations in structure were permitted without losing inhibitory activity; this led to the scheme shown in Fig. 2, where the probable interactions of SAH with the active site of the methylases are shown; this does not differ very much from the scheme proposed by Zappia et al. [*30*] for SAM. It can be seen that amongst the essential chemical functions thought to interact with the enzyme, are the alpha amino group and the carboxyl of homocysteine;

Table 6. Inhibition of P. falciparum in vitro by S-isobutyl-adenosine and two of its analogues [*69*].

Drug	Concentration μM	Parasites per 10,000 red blood cells after 2 days [a]
None	–	172, 225, 206. Av: 201
SIBA	30	261, 237, 191. Av: 230
	100	93, 245, 109. Av: 136
	300	17, 25, 23. Av: 22
None	–	190, 120, 176. Av: 162
SIBA	300	53, 60, 26. Av: 46
	180	146, 140, 139. Av: 139
Iso-SIBA	300	73, 74, 91. Av: 76
	180	128, 146, 202. Av: 158
2-pyridyl-S-adenosine	300	35, 57, 92. Av: 61
	180	122, 162, 199. Av: 161

[a] A petri dish culture was diluted with fresh erythrocytes to give an initial count of 10 parasites per 10,000 red cells and all dishes were prepared from this suspension. They were kept in medium without drug for two days. On days 2 and 3, when the medium was changed the controls were again given medium without drug, whereas the experimental dishes were given medium with drug incorporated in it to give the indicated concentrations. Slides for counts were made on day 4, and hence after two days of exposure to drug.

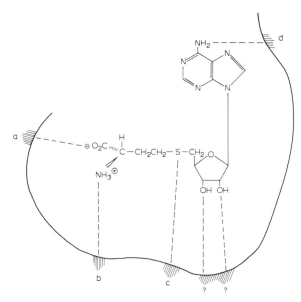

Fig. 2. Proposed enzymatic binding sites for SAH [57].

these are absent in most of the inhibitors we have tested and especially in SIBA and iso-SIBA.

The precise mechanism by which these analogues inhibit oncogenic cell transformation is not yet known, but the effect of SIBA on some biosynthetic events in normal and in virus-infected cells has been studied in detail.

Effect on methylation of nucleic acids. Our preliminary results concerning methylation of tRNA in whole, normal and transformed chicken fibroblasts indicate an increase of the concentration of the following methylated bases upon transformation: 7 MeG (+53%), 3 MeC (+64%), 6 MeA (+100%), 5 MeU (+98%) (M. Vedel, [XII]). When infected cells are cultivated in the presence of 0.5 mM SIBA the increase in the concentration of 6 MeA and of 5 MeU is no more observed. The methylation of the two other bases is not inhibited. When the methylation pattern of tRNA of normal cells grown in the absence and in the presence of 0.5 mM SIBA was compared, we found a 60% decrease in the concentration of 2 Me_2G in the treated cells. Jacquemont and Huppert [67] showed that SIBA inhibits methylation of viral mRNA: 1 mM SIBA inhibits reversibly Herpes simplex virus production in human epidermoid carcinoma cells by blocking the methylation, and especially the capping at the 5' end of viral mRNA. The effect of SIBA on methylation of DNA, tRNA and mRNA in vivo is now under study in our laboratory (M. Vedel, unpublished).

Effect on methylation of proteins. SIBA and all the analogues described here are better inhibitors of protein methylase I (which methylates arginine residues in proteins [6,7]) than of tRNA methylases, at least in vitro (Table 1) (J. Enouf, unpublished); all these compounds are competitive inhibitors of SAM. A strong inhibition of protein methylase I of mouse ascites cells by some of these analogues was also observed by P. Casellas and P. Jeanteur (Montpellier, [VI]).

Effect on incorporation of leucine, thymidine and uridine. We have found that 1 mM SIBA inhibits [^{3}H]-leucine incorporation into proteins by 60% in normal and by 80% in RSV-transformed chick embryo fibroblasts. This effect on transformed cells remains, even after removal of the inhibitor from the medium. The inhibitory effect of SIBA on nucleic acid synthesis is somewhat more difficult to study, as this molecule strongly interferes with the uptake of radioactive nucleic acid precursors by the cells. This effect of SIBA and some of its analogues on cell membranes of normal and transformed cells is now under study (A. Pierré, unpublished).

It seems, however, that DNA synthesis is slightly inhibited (32%) in normal cells treated with 0.05 mM SIBA. At this concentration the uptake of labelled thymidine is inhibited only by 12%, whereas uridine uptake is inhibited by 54%.

Toxicity and intracellular metabolism

Toxicity experiments were performed with mice; SIBA was shown to be nontoxic when injected intraperitoneally in doses up to 500 mg/kg (A.M. Robert, personal communication).

The metabolism of SIBA in normal cells has been studied with ^{14}C-labelled SIBA; the compound enters the cell rapidly and is partially deaminated to 5'-deoxy-5'-S-isobutyl-inosine, which is much less active than SIBA. A small part of SIBA is hydrolysed to adenine and S-isobutyl-ribose (F. Lawrence, [V]).

Conclusion. We have shown that some synthetic analogues of S-adenosylhomocysteine and in particular SIBA are more or less active inhibitors of the transformation of eukaryotic cells by oncogenic viruses, that they also have an antimitogenic action on mitogen-stimulated lymphocytes and that they also are active against malaria parasites growing in human erythrocytes.

Their mechanism of action has been studied, especially the inhibition of methylation of tRNA and of histones; it is, however, possible that other interactions contribute to their activity in the cells. New analogues with less toxicity, greater activity and greater stability inside the cells will have to be prepared.

It remains to be seen whether these compounds are sufficiently well tolerated by higher organisms to be of clinical value and — especially, if they have a more selective activity than other antiviral adenine derivatives, such as ara-A or 3-deaza-adenosine (G. Cantoni, [XI]).

Acknowledgements

We thank Drs. P. Vigier, L. Montagnier, J.C. Cherman and M. Crepin for stimulating discussions; the work of our laboratory would not have been possible without the competent participation of our junior colleagues: A. Berneman, P. Blanchard, J. Enouf, C. Hedgecock, F. Lawrence, A. Pierré, A. Raies, M. Richou, M. Vedel and M. Vuilhorgne.

Our research was supported, in part, by grants from the Ligue Nationale Française contre le Cancer, the Institut Pasteur, the Centre National de la Recherche Scientifique (ATP 'Pharmacologie des Substances anticancéreuses et immunomodulatrices'), the Délégation Générale à la Recherche Scientifique et Technique ('Cancérogénèse et Pharmacologie du Cancer') and a grant from the C.E.A. (Commissariat à l'Energie Atomique, Saclay) for purchase of labeled compounds.

References

1. Du Vigneaud V., Chandler J.P., Moyen A.W., Keppel D.M. (1939) J. Biol. Chem. *131*, 57.
2. Cantoni G.L. (1952) J. Am. Chem. Soc. *74*, 2942.
3. Lederer E. (1977) In: Salvatore F., Borek E., Zappia V., Williams-Ashman H.G., Schlenk F., eds.: *The Biochemistry of Adenosylmethionine*. Columbia University Press, New York, N.Y., p. 89.
4. Starr J.L., Sells B.N. (1969) Physiol. Rev. *49*, 623.
5. Kerr S.J., Borek E. (1972) Adv. Enzymol. Relat. Subj. Biochem. *36*, 1.
6. Paik W.K., Kim S. (1971) Science *174*, 114.
7. Paik W.K., Kim S. (1975) Adv. Enzymol. Relat. Subj. Biochem. *42*, 227.
8. Ballou C.E. (1977) In: Salvatore F., Borek E., Zappia V., Williams-Ashman H.G., Schlenk F., eds.: *The Biochemistry of Adenosylmethionine*. Columbia University Press, New York, N.Y., p. 435.
9. Salvatore F., Cimino F. (1977) In: Salvatore F., Borek E., Zappia V.,

Williams-Ashman H.G., Schlenk F., eds.: *The Biochemistry of Adenosylmethionine.* Columbia University Press, New York, N.Y. p. 187.
10. Nau F. (1976) Biochimie *58*, 629.
11. Nau F. (1977) In: Salvatore F., Borek E., Zappia V., Williams-Ashman H.G., Schlenk F., eds.: *The Biochemistry of Adenosylmethionine.* Columbia University Press, New York, N.Y. p. 258.
12. Furuichi Y., Muthukrishnan S., Tomasz J., Shatkin A.J. (1976) Prog. Nucleic Acid Res. Mol. Biol. *19*, 3.
13. Colonno R.J., Abraham G., Banerjee A.K. (1976) Prog. Nucleic Acid Res. Mol. Biol. *19*, 83.
14. Shatkin A.J. (1976) Cell *9*, 645.
15. Klootwijk J., Planta R.J., Bakker J. (1976) J. Microsc. Biol. Cell *26*, 91.
16. Kudryashova I.B., Kirnos M.D., Vanyushin B.F. (1976) Biokhimiya *41*, 1968.
17. Cato A.C.B., Burdon R.H. (1977) Biochem. Soc. Trans. *5*, 675.
18. Browne M.J., Burdon R.H. (1977) Nucleic Acids Res. *4*, 1025.
19. Hadi S.M., Bickle Th.A., Yuan R. (1977) J. Biol. Chem. *250*, 4159.
20. Linn S., Lautenberger J.A., Eskin B., Lackey D. (1977) Fed. Proc. Fed. Am. Soc. Exp. Biol. *33*, 1128.
21. Berkner K.L., Folk W.R. (1977) J. Biol. Chem. *252*, 3185.
22. Szmelcman S., Adler J. (1976) Proc. Natl. Acad. Sci. USA *73*, 4387.
23. Springer M.S., Goy M.F., Adler J. (1977) Proc. Natl. Acad. Sci. USA *74*, 3312.
24. Aswad D., Koshland D.E., Jr. (1975) J. Mol. Biol. *97*, 207.
25. Springer W.R., Koshland D.E., Jr. (1977) Proc. Natl. Acad. Sci. USA *74*, 533.
26. Srinivasan P.R., Borek E. (1964) Science *145*, 548.
27. Mandel L.R., Hacker B., Maag T.A. (1977) Cancer Res. *29*, 2229.
28. Kerr S.J. (1977) Cancer Res. *35*, 2969.
29. Deguchi T., Barchas J. (1971) J. Biol. Chem. *246*, 3175.
30. Zappia V., Schlenk F., Zydek-Cwick C.R. (1969) J. Biol. Chem. *244*, 4499.
31. Razin A., Goren D., Friedman J. (1975) Nucleic Acids Res. *2*, 1967.
32. Kerr S.J. (1972) J. Biol. Chem. *247*, 4248.
33. Forrester P.I., Hancock T.L. (1976) Cancer Lett. *1*, 161.
34. Wainfan E., Landsberg B. (1971) FEBS Lett. *19*, 144.
35. Tekitek A., Berreur-Bonnenfant J., Rojas M., Ferezou J.P., Barbier M., Lederer E. (1977) FEBS Lett. *80*, 348.
36. Tekitek A., Rojas M., Berreur-Bonnenfant J., Pham G., Nau F., Ferezou J.P., Barbier M., Lederer E. (1977) C.R. Acad. Sci. *285*, Ser. D 825.
37. Sen, G.C., Lebleu B., Brown, G.E., Rebello M.A., Furuichi Y., Morgan M., Shatkin A.J., Lengyel P. (1975) Biochem. Biophys. Res. Commun. *67*, 427.
38. Lebleu B., Sen G., Shaila S., Lengyel P. (1977) Arch. Int. Physiol. Biochim. *85*, 179.
39. Lutz W.B., Creveling C.R., Daly J.W., Witkop B. (1972) J. Med. Chem. *15*, 795.
40. Chimura H., Sawa T., Takita T., Matsuzaki M., Takeuchi T., Nagatsu T., Umezawa H. (1973) J. Antibiot. *26*, 112.

41. Chimura H., Sawa T., Kumada Y., Nakamura F., Matsuzaki M., Takita T., Takeuchi T., Umezawa H. (1973) J. Antibiot. *26*, 618.
42. Fuller R.W., Molloy B.B., Day W.A., Roush B.W., Marsh M.M. (1973) J. Med. Chem. *16*, 101.
43. Fuller R.W., Roush B.W., Snoddy M.D., Day W.A., Molloy B.B. (1975) J. Med. Chem. *18*, 304.
44. Jensen N.P., Schmitt S.M., Windholz T.B., Shen T.Y., Mandel L.R., Lopez-Ramos B., Porter C.C. (1972) J. Med. Chem. *15*, 341.
45. Avruch L., Fischer S., Pierce H., Jr., Oehlschlager A.C. (1976) Can. J. Biochem. *54*, 657.
46. Borchardt R.T., Wu Y.S. (1974) J. Med. Chem. *17*, 862.
47. Borchardt R.T., Huber J.A., Wu Y.S. (1974) J. Med. Chem. *17*, 868.
48. Borchardt R.T., Wu Y.S. (1975) J. Med. Chem. *18*, 300.
49. Borchardt R.T., Huber J.A., Wu Y.S. (1976) J. Org. Chem. *41*, 565.
50. Coward J.K., Sweet W.D. (1972) J. Med. Chem. *15*, 381.
51. Coward J.K., Bussolotti D.L., Chang C.D. (1974) J. Med. Chem. *17*, 1286.
52. Chang C.D., Coward J.K. (1975) Mol. Pharmacol. *11*, 701.
53. Chang C.D., Coward J.K. (1976) J. Med. Chem. *19*, 684.
54. Kogan M.V., Venkstern T.V., Rekunova V.N., Yurkevich A.M., Baev A.A. (1976) Mol. Biol. *10*, 73.
55. Gnegy M.E., Lotspeich F.J. (1976) J. Med. Chem. *19*, 1191.
56. Montgomery J.A., Shortnacy A.T., Thomas H.J. (1974) J. Med. Chem. *17*, 1197.
57. Borchardt R.T. (1977) In: Salvatore F., Borek E., Zappia V., Williams-Ashman H.G., Schlenk F., eds.: *The Biochemistry of Adenosylmethionine*. Columbia University Press, New York, N.Y., p. 151.
58. Hildesheim J., Hildesheim R., Lederer E. (1971) Biochimie *53*, 1067; (1972) *54*, 431.
59. Hildesheim J., Hildesheim R., Yon J., Lederer E. (1972) Biochimie *54*, 989.
60. Hildesheim J., Goguillon J.F., Lederer E. (1973) FEBS Lett. *30*, 177.
61. Hildesheim J., Hildesheim R., Blanchard P., Farrugia G., Michelot R. (1973) Biochimie *55*, 541–546.
62. Michelot R., Legraverend M., Farrugia G., Lederer E. (1976) Biochimie *58*, 201.
63. Legraverend M., Michelot R. (1976) Biochimie *58*, 723.
64. Robert-Géro M., Lawrence F., Farrugia G., Berneman A., Blanchard P., Vigier P., Lederer E. (1975) Biochem. Biophys. Res. Commun. *65*, 1242.
65. Pierré A., Richou M., Lawrence F., Robert-Géro M., Vigier P. (1977) Biochem. Biophys. Res. Commun. *76*, 813.
66. Raies A., Lawrence F., Robert-Géro M., Loche M., Cramer R. (1976) FEBS Lett. *72*, 48.
67. Jacquemont B., Huppert J. (1977) J. Virology *22*, 160.
68. Bona C., Robert-Géro M., Lederer E. (1976) Biochem. Biophys. Res. Commun. *70*, 622.
69. Trager W., Robert-Géro M., Lederer E. (1977) FEBS Lett., *85*, 264.
70. Legraverend M., Ibanez S., Blanchard P., Enouf J., Lawrence F., Robert-Géro M., Lederer E. (1977) Eur. J. Med. Chem. Chim. Ther. *12*, 105.

Additional recent references

I. In vivo inhibition of Novikoff cytoplasmic messenger RNA. Methylation by S-tubercidinylhomocysteine. Kaehler M., Coward J., Rottman F. (1977) Biochemistry *16*, No. 26, 5770–5775.
II. S-adenosyl-L-homocysteine hydrolase: analogs of S-adenosyl-homocysteine as potential inhibitors. Chiang P.K., Richards H.H., Cantoni G.L. (1977) J. Mol. Pharmac. *13*, 939–947.
III. Avian oncornavirus associated N^2-methylguanine transferase, location and origin. Pierre A., Berneman A., Vedel M., Robert-Géro M., Vigier P. (1978) Biochem. Biophys. Res. Commun. *81*, 315–321.
IV. DNA Methylase activity associated with Rous sarcoma virus. Berneman A., Robert-Géro M., Vigier P. (1978) FEBS Lett. *89*, 33–36.
V. Identification of some metabolic products of 5'-deoxy-5'-S-isobutylthioadenosine, an inhibitor of virus-induced cell transformation. Lawrence F., Richou M., Vedel M., Farrugia G., Blanchard P., Robert-Géro M. (1978) Eur. J. Biochem. *87*, 257–263.
VI. Protein methylation in animal cells. II. Inhibition of S-adenosyl-L-methionine:protein (arginine-methyltransferase by analogs of S-adenosyl-L-homocysteine. Casellas P., Jeanteur, P. (1978) Biochim. Biophys. Acta *519*, 243–268.
VII. Effect of 5'-deoxy-5'-S-isobutyl-adenosine (SIBA) on mouse mammary tumor cells and on the expression of mouse mammary tumor virus. Terrioux C., Crepin M., Gros F., Robert-Géro M., Lederer E. (1978) Biochem. Biophys. Res. Commun. *83*, 673–678.
VIII. S-Adenosylhomocysteine analogues as inhibitors of specific tRNA methylation. Leboy Phoebe S., Glick Jane M., Steiner Fresia S., Haney Sharon, Borchardt Ronald T. (1978) Biochim. Biophys. Acta *520*, 153–163.
IX. Sinefungin, a potent inhibitor of virion mRNA (guanine-7-)-methyltransferase, mRNA (nucleoside-2'-)-methyltransferase, and viral multiplication. Pugh Charles S.G., Borchardt R.T., Stone Henry O. (1978) J. Biol. Chem. *253*, 4075–4077.
X. Cordycepin and xylosyladenine: inhibitors of methylation of nuclear RNA. Glazer Robert I., Peale Ann L. (1978) Biochem. Biophys. Res. Commun. *81*, 521–526.
XI. 3-Deazaadenosine an inhibitor of adenosylhomocysteine hydrolase inhibits reproduction of Rous sarcoma virus and transformation of chick embryo cells. Bader John P., Brown Nancy R., Chiang Peter K., Cantoni Giulio L. (1978) Virology *89*, 494–505.
XII. Inhibition of tRNA methylation in vitro and in whole cells by an oncostatic S-adenosyl-homocysteine (SAH) analogue: 5'-deoxy-5'-S-isobutyl-adenosine (SIBA). Vedel M., Robert-Géro M., Legraverend M., Lawrence F., Lederer E. (1978) Nucleic Acids Research *5*, 2979–2989.

CHAPTER 8

Ion transport in membranes and the ionophore problem

Yu.A. OVCHINNIKOV

Among the basic approaches to the structural and dynamic characteristics of membranes, a prominent place is occupied by problems involving ion transport. This is quite understandable, for ion transport plays one of the key roles in the functioning of bioenergetic and receptor systems, in the transmission of nerve impulses and in other processes in the organism. The recent achievements herein are the climax of a number of fundamental discoveries made possible through the development of reliable and accurate methods, and the creation of improved technical means [1].

If one casts a general view on the study of ion transport in biomembranes he can clearly discern two main trends.

One is the direct tracing of ion transport in multifarious biological membranes. Although, because of the complexity of membrane structures and processes, such an approach seemed to hold little promise, our advancing knowledge of membranes to new, higher levels, should lead one to expect from it more tangible results.

Indeed, the classical concept of a biomembrane based on the Danielli-Davson model [2,3] has evolved through a series of modifications [4–6] to the presently accepted dynamic mosaic model [7–9]. Very important results are being garnered from studies of membrane lipids [10–12]. An impetus has finally been given to the study of membrane proteins and lipoproteins [13–15] which had long been at a standstill. This has resulted, among other findings, in the observation that under appropriate conditions true membrane proteins can often be isolated in a homogeneous state and characterized both functionally and chemically. In the elucidation of the nature of biological membranes, including the interrelationships of their individual components, besides electron microscopy, novel, particularly physical methods of study [16–22] have acquired more and more importance. Of considerable import to the understanding of ion transport processes was the detailed biochemical study of membrane mechanisms associated with the generation and transformation of energy [23–30]. Mention should be made of the elegant experiments on the reassembly of biomembrane transport systems [31–35] and also on the use of normal and inside-out vesicles in bacterial membrane investigations [36–38]. All such events have paved the way to the direct study of ion-transport systems in diverse biological membranes.

The second direction of transmembrane ion transport studies involves the use of model systems simulating biological membranes in certain parameters. Such an approach is particularly effective if one can properly correlate the results on artificial membranes with those on biological membranes, and, thereby, take full advantage of the ease of experimenting with the relatively simple and easily controllable model systems. A major step in this area of membranology was the widespread use of artificial phospholipid systems — bilayers [39–41] and, subsequently, liposomes [42,43]; later, other types of artificial membranes were developed [44]. It is noteworthy, that practically at the same time a number of substances of microbial origin were found to be capable of selectively increasing the membrane ion permeability by forming a lipophilic complex with a given ion and transporting this ion in the complexed form across the membrane [45–47]. Such substances of relatively low molecular weight and their synthetic analogs have received the name of ionophores [48,49]. Closely related microbial products have been found that model ion channels in biomembranes, increasing the ion permeability of artificial membranes by transmitting the ions through channels they form [47,50–57]. The presence of ionophores and similar substances in artificial membrane systems endowed the latter with many of the basic characteristics of their biological counterparts, in particular with translocating efficiency and selectivity [58,59]. As a result, model membranes have become powerful tools in the study of ion transport in the cell.

In a comparatively short time many basic problems in the induced transport of ions in model membranes and, in some cases in biological membranes, were solved [60–62], new types of naturally occurring and synthetic ionophores were discovered [1] and the mode of their action was elucidated in detail [63–69].

The intensive and extensive study of ionophores and their behavior in membrane systems immensely enriched our knowledge of the major pathways and concrete mechanisms of selective transmembrane ion transport, and laid a firm basis for further study in this direction. It seemed reasonable to expect that Nature makes use of similar means for the transport of ions in cellular membranes, and the search for naturally occurring ionophores in cells was not only highly enticing but appeared to be not without rational grounds. However, at present they can hardly be justified, although some workers are still persisting in this direction [70–73]. At the same time it has become clear that, as a rule, ion transfer in biological membranes is effected by quite complicated protein systems, in all probability acting largely in the form of ion-transmitting channels. Such systems were initially postulated on theoretical grounds, for nervous membranes, and were later detected in other types of biomembranes by electrophysiological and electrochemical methods. In very recent time attempts have been made to directly identify the transmitting complexes or their composite parts, including to isolate them by biochemical means and carry out their structural study. At the present state of art it would be very difficult to give an unequivocal answer to the question of how the ion-conducting systems of biomembranes are formed, what is the actual physicochemical mechanism underlying ion transport across biological membranes, and whether there is a single mode of such transport common to all membranes, but we have every reason to believe that we are quite close to such an answer.

The present review has as its objective a brief discussion of the present state of our knowledge of ion transport in membrane systems and a summary of the work which appeared after publication of our monograph [1], without, however, any pretense to comprehensiveness.

Ionophores

The idea that lipid-soluble ion complexes are involved in membrane electrolyte transport was expressed in a general way as far back as half a century ago and afterwards repeatedly discussed by a number of workers [74,75]. However, in its present aspects the problem was formulated in the middle of the sixties by Pressman, who found that a series of naturally occurring substances, primarily antibiotics, affected selective ion transport in mitochondria, and proposed for these compounds the name 'ionophores' [45]. At present, hundreds of such compounds, both naturally occurring and synthetic are known [1].

By ionophores one understands substances of relatively low molecular weight, capable of directly transporting ions across artificial and biological membranes in the form of lipophilic complexes (Fig. 1). Thus, a major characteristic of a ionophore is its ability to complex the ion it is transporting. Such complexing behavior was observed in a number of laboratories very shortly after discovery of the transporting activity. It is, by the way, just this property which in our opinion distinguishes the true ionophores from other systems involved in the transmembrane ion transporting process. Practically all the ionophores known heretofore are members of a new type of complexones which, through their polar groups, bind the ion by means of ion-dipole interaction. The number and

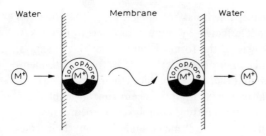

Fig. 1. Schematic of the mode of ionophore action.

the spatial arrangement of ligands binding the ion obey the principles of coordination chemistry, and are specific to the given ion. It should be noted that the complexing properties of the ionophores are of wider significance than their ionophore activity. With their discovery, a new chapter in modern chemistry has been opened, and their properties are being utilized — applied chemistry, technology and medicine. Thus, they are serving as the basis for a new type of electrodes, as extracting agents for concentrating pure elements, as homogeneous catalysts, etc. [1,66].

Another important characteristic of ionophores is the complexing-determined selectivity of their mediation of ion transport. This selectivity has its origin in the molecular structure of the ionophore in which a decisive factor is not so much the purely geometric parameters of the molecule as its conformational dynamics [76]. The majority of ionophores possess cyclic or pseudocyclic structures, the complexed ion being enclosed in the molecular cavity. Often such highly selective ion binding is achieved through the formation of 'sandwiches' or similar structures with a complexone : ion ratio of 2 : 1, 3 : 2 etc. [77]. It should be noted that the selectivity is also highly dependent on the hydration/dehydration energy of the ions, inasmuch as the ion in the ionophore complex is partially or completely desolvated [1,76]. Mention might be made that it is just these ionophores which have provided us with the most dramatic examples of selectivity, particularly regarding K^+ and Ca^{2+} ions [1,49,66].

A third major property of the ionophores is their lipophilicity, which is due not only (and not always) to the free molecules, but, more significantly, to their complexes with the transported ion. It often results from the specific spatial arrangement of the molecule, which 'embraces' as it were, the ion with its polar 'arms', exposing to the external medium the fatty hydrocarbon radicals. Such molecular packing of the complex creates the conditions for its ready penetration into the membrane's lipid phase (lipid, organic solvent etc.) and for its subsequent movements therein (see Fig. 1).

Finally, a fourth criterion of a ionophore is its ability to comply with the subtle requirements of the kinetic parameters in every stage of the process. Firstly, the translational velocity in the inner regions of the membrane must be of sufficient magnitude to provide for the efficiency of the transport process (of the order of 10^4-10^5 ions/sec). At the same the the ion-binding constant must be such that the ionophore could effectively bind the ion on one side of the membrane and easily give it up on the other side. Hence, a strong complexone may not always be a good ionophore; and complexone properties alone are insufficient for the manifestation of ionophorous activity — a good ionophore must possess a number of other special properties.

Meriting special consideration is the important and intriguing problem of the transmembrane transport of hydrogen ions. The vital importance to the living cell of these highly specific charged entities is a matter of ancient knowledge, and the proton gradient in membranes is one of the cornerstones of bioenergetics.

Not surprising, therefore, is the attention given to the question of how protons are transported across a membrane. Do there exist, for example, special proton ionophores, protonophores, so to say? It is easy to conceive them theoretically, but are they actually operating in biological membranes, or are there other mechanisms by which such translocation actually occurs, such as passage through channels?

In recent years extensive use is being made of synthetic protonophores as tools in membrane study. Such substances belong to various chemical classes, but, as a rule, they are weak acids. Among them are 2,4-dinitrophenol (DNP), pentachlorophenol (PCP), 4,5,6,7-tetrachlorosalicylanilide (TCSA), dicoumarol (DC), 2-trifluoro-methyl-4,5,6,7-tetrachlorobenzimidazole (TTCB), carbonylcyanide *m*-chlorophenylhydrazone (CCCP), carbonyl cyanide *p*-trifluoro-methoxyphenylhydroazone (CCFP), 1,2-dicarbodecachlorododecaborane (DCCB), etc.; one may add here also substances of natural origin ending with antibiotics and alkaloids. Being lipophilic, these agents readily penetrate the membrane in undissociated form, giving off a proton on the opposite side, and returning as the respective anion. In this way they play the role of efficient uncouplers of oxidative phosphorylation [78–88]. Although their binding of the transported ion is simpler than in the case of the metal ionophores, and they are thus less selective, they can be considered as full-fledged members of the ionophore family.

Naturally, of considerable interest is whether such protonophores exist in biological membranes. Many authors have discussed likely candidates for this role, but there seems to be no definite answer to this question. There is some likelihood that such mobile proton carriers in mitochondrial and chloroplast membranes might be certain links in the redox chain, in particular ubiquinones and plastoquinones [89–92]. The probability that this is so becomes higher if it be remembered that these lipophilic quinones readily transport H atoms, and that the necessary balance of charge can be maintained by parallel transport of charged ions or electrons (see [93]).

Quite clearly, ionophores are capable of increasing not only the permeability of artificial membranes, but also passive ion transport across biological membranes. Depending upon whether the ion-bearing ionophore is charged or neutral (the latter case occurring when the free ionophore itself carries a charge opposite to that of the ion), the ion transport proceeds along a potential or a concentration gradient. It is this gradient which is the driving force of the process. At the same time, when present in biological membranes, ionophores exert a strong influence on ion flow and on the general ion equilibrium, thereby having an indirect effect on activated ion transport. Then, again, one may theoretically conceive of a system in which the ionophore is part of a transport mechanism coupled to an energy transducer, and in that way can directly monitor the active transport process. No such system has, however, as yet been discovered, despite repeated prediction of its existence.

There are substantial grounds for assuming that ionophores carry the ions across the lipid phase of biological membranes, i.e. across their bilayers or channels, without becoming involved in specific interaction with their protein or carbohydrate components. In particular, this is evidenced by the quite satisfactory correlation between the behavior of many ionophores in biological and various artificial membranes, especially when the lipid composition of the latter is similar to that of the former. Also no confirmation as yet exists of the initial assumption of the occurrence of specific ionophore receptors in biomembranes [46,47]; on the contrary, it appears to be strongly refuted by the 100% ionophorous activity in mitochondria of the total enantiomers of a number of naturally occurring ionophores [94,95].

In a general way, the mechanism of ionophore-mediated ion transport has been considerably elucidated in the case of artificial membranes. However, a number of important aspects of the mechanism still remains obscure. Such are, for instance, the relative importance in each concrete case of, say, direct ion transport by a single ionophore molecule, and ion transmission from one molecule to another by a relay mechanism; in other words, the problem of how important to the ionophorous function is the cooperative principle, and what is the contribution to the overall active transport process of molecular aggregates such as sandwiches, or their like? Finally, we are still unclear as to the specific behavior of the ionophores and their complexes at interfacial boundaries and in the superficial zone of the membrane. These and similar problems are at present undergoing intensive investigation [60–69].

From the above-said it follows that ionophores are very subtle instruments for the control of transmembrane ion transport and serve as good illustration of Nature's filigree designing. It is highly probable that such effective bioregulators required long evolutionary periods for their perfection. Of interest in this light is the recent statement by V. Prelog [96] that, in the process of the origination of life, substances such as ionophores might have served as a certain stage in the development of selectivity, an important prerequisite of living matter. There are thus some grounds to regard ionophores as evolutionary precursors of the more sophisticated and more perfected systems responsible for the ion selectivity in a present day cell. The fact that Nature already has excellent ionophores in her workbasket, of course, makes it all the harder for a bench-worker to try to copy them or to perfect their properties in one way or another. Where such an objective was achieved, it was done so only by careful analysis of the structure-function relationship and by studying numerous synthetic analogs. In this respect a favorable situation arose with the peptide antibiotics, particularly of the valinomycin series, whose parent member was the first to have been assigned the name of 'ionophore'.

Valinomycin. Valinomycin was isolated by Brockmann [97] in 1955 and its structure as a cyclododecadepsipeptide (Fig. 2) was finally established in 1963 by Shemyakin and coworkers [98]. Simultaneously we had synthesized very many of its analogs, among them compounds of high biological activity, and

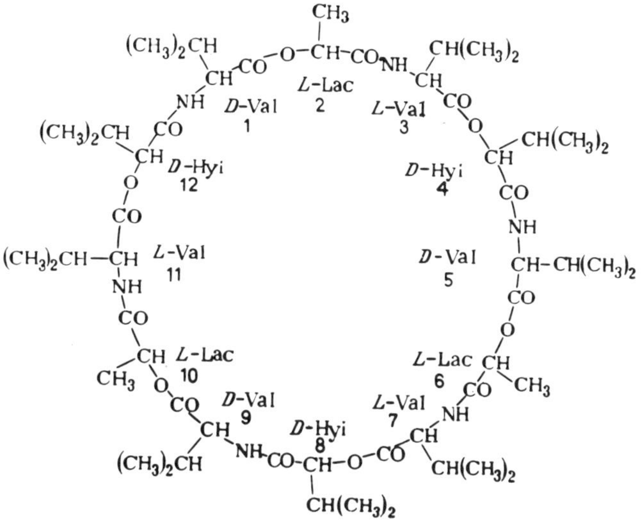

Fig. 2. Valinomycin.

were, therefore, able to obtain the first correlation in this series between structure and biological function [95,99–101]. Although the fact that valinomycin is a potent decoupler of oxidative phosphorylation in mitochondria had been known as early as 1959 [102], its K^+-ionophorous activity was first noted by Pressman in 1964 [45]. From that time on valinomycin has become one of the most popular tools and objects of study among biochemists and biophysicists. It was just after the example of valinomycin that the fundamental principles of ionophorous function were formulated and the physicochemical basis of their action as complexones and transmembrane ion carriers was unraveled.

The mode of discrimination between Na^+ and K^+ ions by valinomycin, originating from its unique spatial structure, was established for the first time in our laboratory by the composite use of a number of physicochemical methods to investigate the dynamic conformational characteristics of this molecule in solution [103]. A little later, the results obtained were confirmed in general terms by an independent X-ray analysis of crystalline samples of the antibiotic [104]. As a result of these studies, it was found that free valinomycin is in a 'bracelet' type of conformation and that this conformation is basically retained in its K^+ complex (Fig. 3) [76]. It is noteworthy, that valinomycin was the first naturally occurring peptide whose solution conformation was elucidated by a composite physicochemical approach in which NMR spectroscopy played a leading role [105]. Subsequent studies of valinomycin and its analogs, using this approach [106–114], and also supplementary X-ray data [115–119], deepened our understanding of the nature of this unique molecule and made possible the predetermined synthesis of many valinomycin-like ionophores with valuable properties [120–125].

Although valinomycin contains only 12 hydroxy and amino acid residues, it

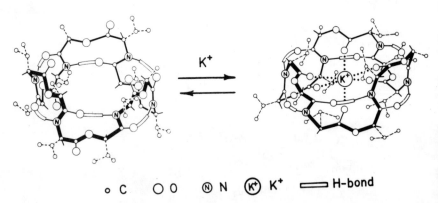

Fig. 3. 'Bracelet' conformation of valinomycin in solution and in the K^+-complex.

is characterized by an amazing versatility of physicochemical and biological properties, usually inherent in much more complicated systems. It woud not be too farfetched to seek a resemblance of this elegant molecule to an enzyme. Indeed, like an enzyme, valinomycin sharply accelerates a certain process, in this case transmembrane ion transport. Its action is manifested in catalytic concentrations ($10^{-9}-10^{-11}$ mole/l). Valinomycin has an amazing substrate specificity, namely to K^+ ions, its selectivity for this substrate being 10,000 times higher than for the Na^+. Moreover, in binding this 'substrate', valinomycin undergoes a 'substrate'-induced conformational 'adaptation', whereby, in order to bind the cation, the ester carbonyls turn from an 'outward' to an 'inward' orientation (see Fig. 2).

It is also to be noted that in the free state valinomycin manifests high surface activity, whereas in the complexed state it is extraordinarily hydrophobic — properties responsible for the excellence of its membrane-affecting activity, and at the same time typical of many membrane enzymes. Finally, valinomycin is a potent antimicrobial agent, a fact fully exploited when correlating the properties of valinomycin and its analogs in various *in vitro* and *in vivo* experiments.

It was the striking ionophorous property of valinomycin that served as stimulus for the search for such compounds in biological membranes and that was the decisive factor in the origin and spread of the entire concept of 'ionophore transport' in modern membranology. We may with full justification state that the study of valinomycin, among other ionophores the closest in nature to proteins, was the first example of successful intervention into the highly complicated area of the molecular basis of membrane ion transport and that it greatly deepened our knowledge of the possible ways that ions could be transported in biological systems.

One should also add that the information accumulated in the study of valinomycin has served as basis for the directed synthesis of analogs with interesting properties which are absent in the parent compound. Thus the K^+ complexes of *cyclo*[-(D-Val-Hyi-Val-D-Hyi)$_3$-] [*126*] and *cyclo*[-(D-Val-MeAla-Val-D-Hyi)$_3$-] [*127*] possess stability constants exceeding by 1–2 orders of magnitude that of the valinomycin complex. When *cyclo*[-(D-Val-Pro-Val-D-Pro)$_3$-] functions as a ionophore, the ion-binding process occurs in the aqueous solution rather than at the membrane-water interface as in the case of valinomycin shown in Fig. 1. It thus must have a different mode of action from valinomycin [*69,128*].

Finally, *cyclo*[-(D-Val-Lac-Val-D-Pro)$_3$-] forms not only 1 : 1 complexes, but also 2 : 1 'sandwiches' [*129*], whereas *cyclo*[-(D-Val-Lac-D-Hyi)$_4$-], comprising four tetradepsipeptide units instead of the three in valinomycin, is an excellent ionophore for bulky organic cations such as trimethylammonium, choline and acetylcholine [*130*].

Enniatins. This group of antibiotics, closely related to valinomycin, was first isolated in 1947 [*131,132*] and their total synthesis-confirmed structure as cyclohexadepsipeptides (Fig. 4) was established in our laboratory in 1963 [*133–135*]. It was in the course of the subsequent study of enniatins A and B and their analogs that the first correlations were obtained between their structure and biological function [*99*], and then their ionophore effect on mitochondria was discovered [*45*]. Soon after, Mueller and Rudin [*58*] suggested the possible existence of equimolar complexes of enniatin B with alkali ions, and such a complex was, in fact, demonstrated experimentally in the crystalline state [*136*] and in solution [*137*]. The K^+-complex of enniatin B (Fig. 5) is in the form if a disc in which, contrary to valinomycin, the centrally located cation is little screened from the solvent. This is the reason for the low stability of the complexes in general, and for the much lower complexing selectivity than that of valinomycin. For instance, the enniatins are capable of binding not only alkali, but also alkaline earth and transition metals [*138,139*]. Quite recently, an important feature of ionophore behavior was discovered in our laboratory on the example of the enniatins, when it was shown that they are capable of forming higher (2 : 1, 3 : 2, etc. ionophore : cation) complexes in both solution and membrane, and that they probably transport ions across membranes in just such a form (Fig. 6). Subsequently, ion transport in the form of 'sandwiches' was

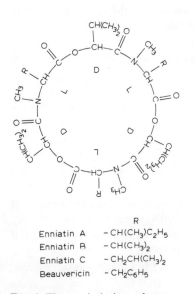

	R
Enniatin A	$-CH(CH_3)C_2H_5$
Enniatin B	$-CH(CH_3)_2$
Enniatin C	$-CH_2CH(CH_3)_2$
Beauvericin	$-CH_2C_6H_5$

Fig. 4. The enniatin ionophores.

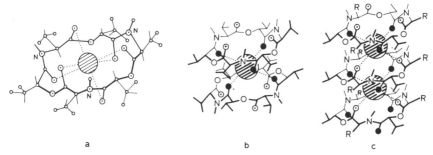

Fig. 5. Conformation of enniatin complexes with a macrocycle : ion ratio of (a) 1 : 1, (b) 2 ; 1, and (c) 3 : 1.

revealed in other ionophores, and it is very likely that such 'sandwich'-mediated transport is responsible for the high ion selectivity of the process [77,140–142].

Considerable data have now been accumulated on the ionophore activity of enniatin, the closely related beauvericin and of their analogs, as well as on the spatial structure of these compounds and of their cation complexes [143,144].

Antamanide. This cyclopeptide (Fig. 7), isolated by Wieland [145–149] from the green mushroom, is a potent antagonist of its principal toxins, amanitine and phalloidine, and in many respects closely resembles peptide ionophores. In fact, Pressman [150] discovered antamanide-induced K^+ transport in mitochondria as

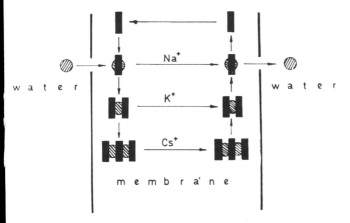

Fig. 6. Model of the enniatin-ionophore transport of alkali metal ions across a lipid membrane.

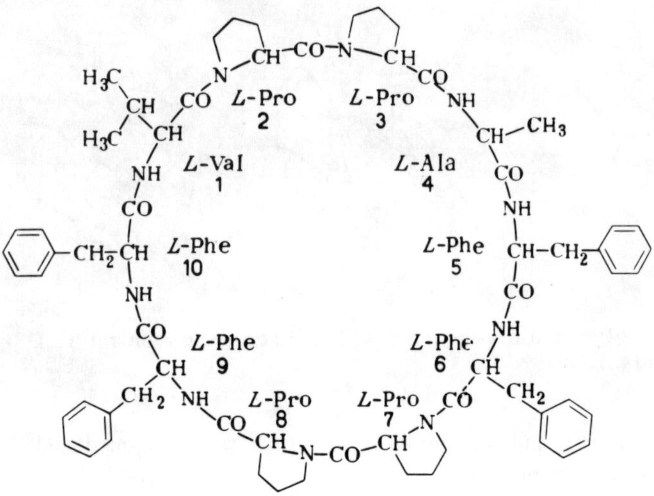

Fig. 7. Antamanide.

far back as in 1965; the effect, however, being much smaller than that of valinomycin.

In a joint work of Wieland's and our laboratories [146], antamanide was found to form metal ion complexes in solution with a striking preference for Na^+ and Ca^{2+}. The solution structure of these complexes was studied [151–154] and, later, a much more detailed determination was made of the crystalline structure [155–160]. The structure of the Li^+-complex of antamanide is shown in Fig. 8 [155]. One could have expected that Na^+ and Ca^{2+} ionophores could have been

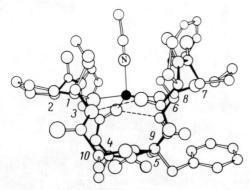

Fig. 8. Conformation of the crystalline Li^+-antamanide complex.

made by modification of antamanide, but none of its numerous synthetic analogs have displayed sufficient ionophore activity on membranes, although a number of compounds have indeed been synthesized that increased the ion flow of alkali metals in liposomes [161]. Work in this direction is in progress.

To the group of neutral ionophores can be referred the synthetic cyclopeptides obtained in very recent times in a number of laboratories [77,162–167]. Their ring size, as a rule, fluctuates between 12 and 24 atoms; they ordinarily contain proline or N-methylamino acid residues and manifest selectivity for alkali metal ions. Of particular interest is the bicyclic derivative synthesized by Schwyzer [168], formed by two cyclopentapeptides joined by a disulfide bridge and selectively complexing K^+ ions. The behavior on membranes of all the aforementioned cyclopeptides is as yet insufficiently elucidated.

Worthy of note is the synthetic neutral ionophore proposed by V. Simon et al. [169–175] (Fig. 9). It displays marked Ca^{2+} selectivity in artificial and biological membranes.

Macrotetrolides. This group of antibiotics was discovered by Prelog et al. in 1955 [176] and their structure (Fig. 10) was later established in the same laboratory [177–180]. After having uncovered the ionophore activity of nonactin, the Swiss workers obtained its crystalline Na^+ complex in the pure state and determined its spatial structure [181,182]. Interestingly, this was the first case of identification of the complex of a ionophore with a metal ion. Its study

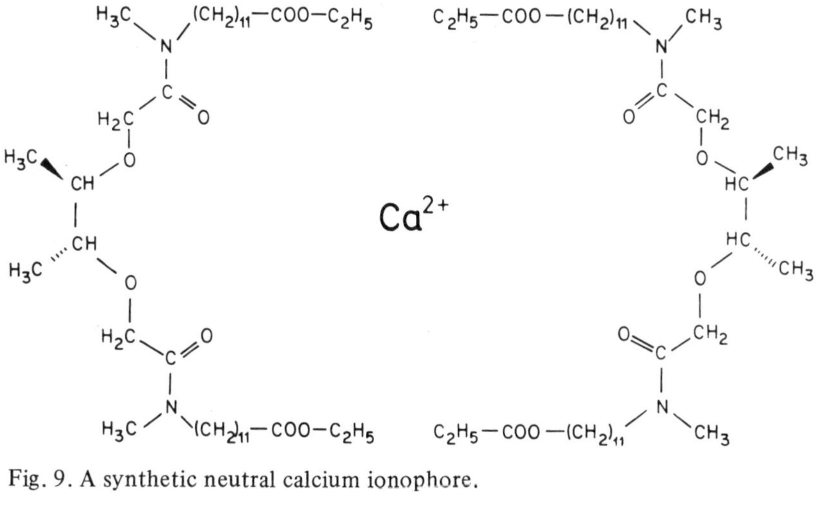

Fig. 9. A synthetic neutral calcium ionophore.

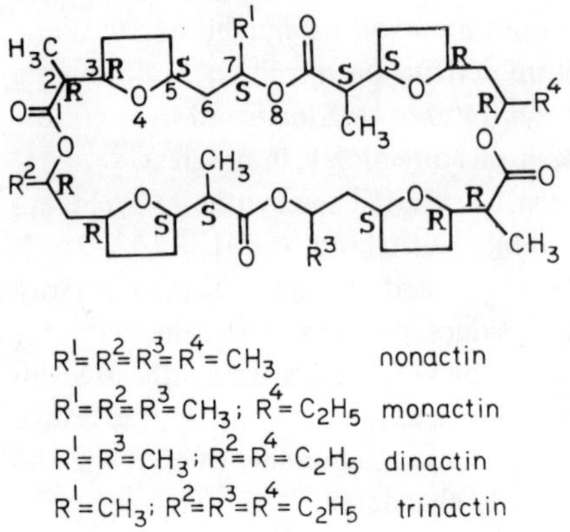

$R^1 = R^2 = R^3 = R^4 = CH_3$ nonactin
$R^1 = R^2 = R^3 = CH_3$; $R^4 = C_2H_5$ monactin
$R^1 = R^3 = CH_3$; $R^2 = R^4 = C_2H_5$ dinactin
$R^1 = CH_3$; $R^2 = R^3 = R^4 = C_2H_5$ trinactin

Fig. 10. Ionophores of the group of nonactins (nactins, macrotetrolides).

resulted in elucidation of the basic principles of the structural organization of such systems. It turned out, in particular, that the ion deprived of its solvate sheath is bound to the antibiotic by ion dipole interaction with the ester carbonyls; the macrotetrolide molecule enveloping, as it were, the cation, assuming the form of the seams of a tennis ball (Fig. 11) [181,182]. Scores of papers have been devoted to the study of the ionophore activity of nactins (see [1]).

Synthetic polyethers.   Cyclic polyethers of the type represented in Fig. 12 do not occur in Nature, they are obtained by organic chemical methods. They were investigated in detail by Pedersen [183–185] and have at present become very

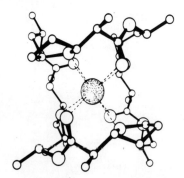

Fig. 11. Conformation of crystalline K^+-nonactin.

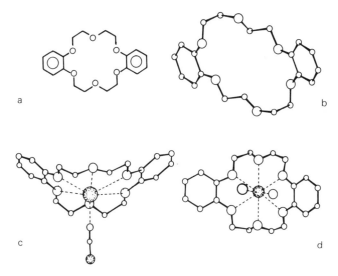

Fig. 12. (a) Dibenzo-18-crown-6, and its conformation, (b) in the crystalline state, (c) in the complex with rubidium thiocyanate, and (d) with sodium bromide.

popular tools for the study of transmembrane ion transport [182,186,187]. Such macrocycles are known as 'crowns' (for instance the compound in Fig. 10 is called dibenzo-18-crown-6). There may be from 9 to 60 atoms including O, S and N in the ring, and the rings may have substituents of highly diverse nature [1,188–192]. Naturally, polyethers differ considerably in cation selectivity, both with respect to complexing and to ionophore activity. Thus, polyethers are known that selectively bind Li^+, Na^+, K^+, Ag^+, Ni^{2+}, Mg^{2+}, Ca^{2+}, Sr^{2+}, Ba^{2+}, Zn^{2+}, Fe^{2+}, Fe^{3+}, Co^{3+}, or other such ions and also substituted ammonium ions [193–201]. Of interest are the asymmetric cyclopolyethers which Cram [202–208] and Prelog [209] have recently synthesized; these compounds possess optical activity ('chiral' ionophores) (Fig. 13), and can selectively bind enantiomers such

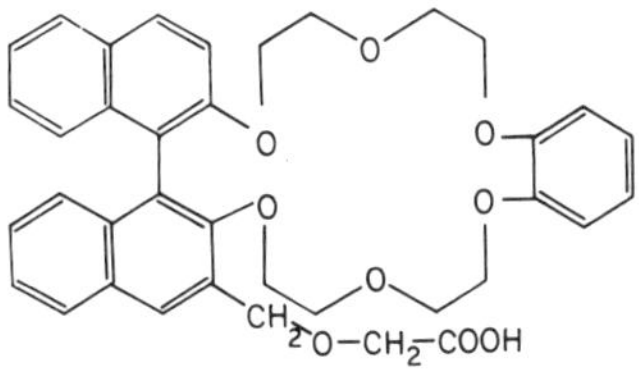

Fig. 13. The chiral cyclopolyether obtained from binaphthol.

as the L- and D-forms of α-phenylethylamine [204–210]. It should also be mentioned that many synthetic polyethers are prone to form 'sandwiches' and other more complicated complexes, particularly with respect to the larger cations [1]. At present cyclopolyethers are utilized not only for membrane study, but also in various areas of chemistry and technology, for conferring organic solubility on metal salts, for the extraction of certain metals [211–220], for regulating the activity of enzymes [221], for activating anions [222–226], etc.

Macrobicyclic polyethers with interesting structural features and properties have been synthesized and studied by Lehn [227–233]. These compounds, of the general formula represented in Fig. 14, are capable of forming stable complexes with highly diverse cations (such complexes have received the name of cryptates). Fig. 15 shows the crystalline spatial structure of such a cryptate carrying a barium ion (cf. Fig. 12, m=n=1, p=2, X=Y=O). Despite their very high complexing selectivity, cryptates cannot function as membrane ionophores because of the exceedingly high complex stability. One may predict, however, that appropriate chemical modification of these polyethers will soon permit them to occupy a fitting position amongst the ionophore family.

Here, another group of compounds of peptide nature which very likely participate in the biological transport of iron ions, may be mentioned. The structure and properties of these compounds known as 'ferrichromes' have been investigated by Neulands [234]. However, they are very far from the true ionophores occurring in Nature.

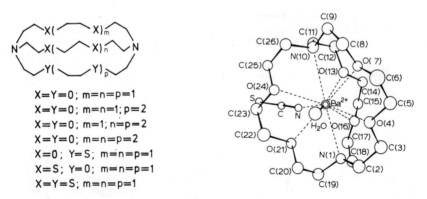

X=Y=O; m=n=p=1
X=Y=O; m=n=1; p=2
X=Y=O; m=1; n=p=2
X=Y=O; m=n=p=2
X=O; Y=S; m=n=p=1
X=S; Y=O; m=n=p=1
X=Y=S; m=n=p=1

Fig. 14. The structure of some macrobicyclic polyethers (cryptands).

Fig. 15. Conformation of the crystalline cryptate ($m = n = 1$, $p = 2$, $X = Y = O$, Fig. 13) with barium thiocyanate.

The group of nigericin. Nigericin, discovered in 1950 [*235,236*] is the parent compound of the 'charged' ionophores which differ in their mode of action in membranes from the neutral (in the free state) ionophores of the type of valinomycin.

After discovery of the ionophore behavior of nigericin in mitochondria [*237*], numerous other such compounds were revealed and the list is ever growing [*1,66,238–252*].

The structure of nigericin (Fig. 16) and its Ag^+ complex (Fig. 17) was established by two research teams in 1968 by X-ray analysis [*253,254*]. A distinguishing feature of the structure, characteristic of most ionophores of this group, is its linearity, the chain consisting mostly of tetrahydrofuran and tetrahydropyran units and carrying a free carboxyl group at the terminus. Although linear, in non-polar media or in the crystalline state nigericin and related substances are in a pseudocyclic form, in which ring closure is effected by a head to tail type of stable hydrogen bond through the mediacy of the carboxyl group. It is in this form, that they bind the ion in a lipophilic complex. The complexing reaction is quite selective. For instance, nigericin exceeds all other ionophores, including valinomycin, in the stability of its K^+-complex (1.5 · 10^5 l/mol in CH_3OH). However, it is inferior to the latter in K/Na selectivity. Another member of this group, nonensin, displays the property of Na^+ selectivity [*70*], rare in a ionophore [*255–257*]. The carboxyl ionophores accelerate transmembrane ion transport according to the $Me^+ \leftrightarrow H^+$ type of exchange, as can be clearly seen from U tube experiments [*66*].

Noteworthy is the very recent synthesis in our laboratory of carboxyl-containing valinomycin and enniatin analogs; under certain conditions they behave like nigericin [*258*]. The study of such compounds with the properties of both neutral and charged ionophores adds new light to our understanding of the mechanism of induced transmembrane ion transport.

Antibiotics X-537A and A-23187. These compounds, so awkwardly named by figures, are also members of the charged ionophore series. The special interest in

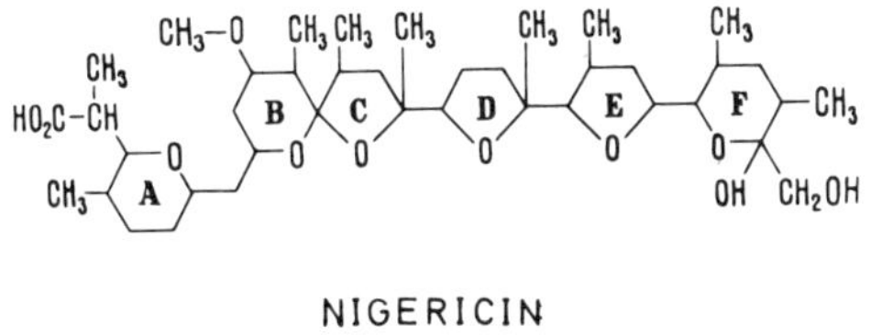

NIGERICIN

Fig. 16. Nigericin.

Fig. 17. Conformation of the crystalline Ag salt of nigericin.

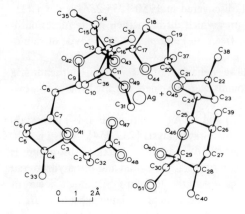

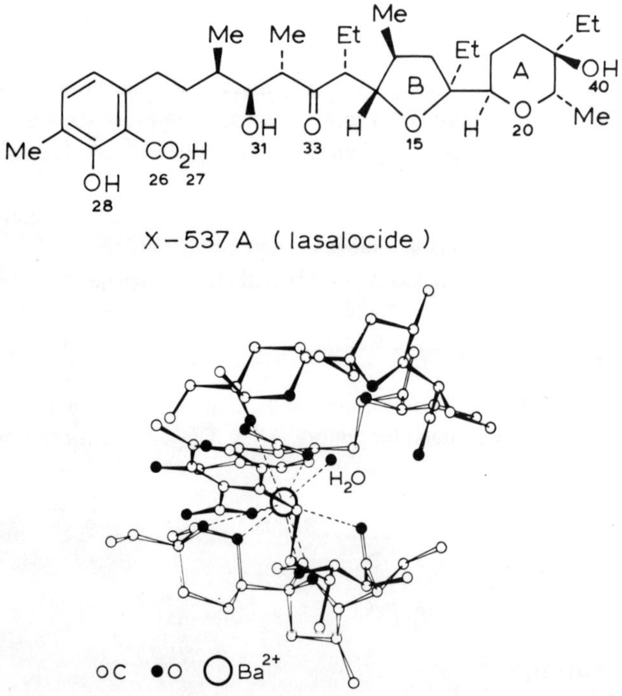

X−537A (lasalocide)

Fig. 18. Lasalocide and its crystalline Ba^{2+}-complex.

them lies in their ability to transport divalent ions, particularly Ca^{2+}, and to affect such processes as muscular contraction, transmission of nerve impulses etc. [*49,66,259*].

Antibiotic X-537A or lasalocide, first isolated in 1951 [*260*], became known as a ionophore only twenty years later [*261*]. Fig. 18 represents the structures of the antibiotic itself and of its 2 : 1 Ba^{2+} complex, respectively [*262*]. Lasalocide is capable of transporting both divalent metal ions ($Ba^{2+} > Sr^{2+} > Ca^{2+} > Mg^{2+}$), and alkali metal ions, as was established by two-phase partition studies and experiments on membranes [*66,259,263*]. The antibiotic X-537A complexes primary amines and also catecholamines [*66,259,264–266*], a property which plays a decisive role in its unique biological action. A number of X-537A analogs have been synthesized, the study of which has shed important light on the structure-function relationship in this group [*261,267–275*].

The divalent-cation-ionophore A-23187, whose action on mitochondria was first observed by Lardy [*276–278*], manifests high Ca^{2+} and Mg^{2+} selectivity. The structure of the antibiotic was elucidated in 1974 [*279*] and recently an X-ray analysis was made of its Ca^{2+} and Mg^{2+} complexes [*280,281*] (Fig. 19).

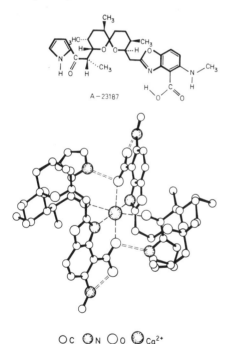

Fig. 19. Antibiotic A-23187 and its crystalline Ca^{2+}-complex.

The complexes have a 2 : 1 composition, which seems to be quite general for carboxylic ionophores.

The behavior of ionophores X-537A and A-23187 has been investigated in detail on many types of artificial and biological membranes [282–289], and lately ways have been indicated for their utilization as potent therapeutic agents in the management of a number of diseases [49,66,259].

Intrinsic ionophores. In studies of transmembrane ion transport, carried out during the last 20 years, it has repeatedly been claimed that certain low molecular membrane components possess the ability of increasing the membrane permeability to metal ions, and sometimes even to manifest selectivity in this process. On these grounds such compounds were ascribed 'ionophorous' activity. Not always could they be regarded as true ionophores, however, and more often than not they had the properties rather of ion channels.

Many investigators are inclined to regard various lipids as likely ionophores. In particular, it could be shown in experiments on artificial membrane systems that, in fact, phosphatidylcholine, phosphatidylethanolamine, phosphatidylserine, sphingomyelin, kephalin, cardiolipin, and other phospholipids selectively increase ion flows in membranes [290–298]; a similar behavior has been noted in some free fatty acids [299,300].

However, direct proof of the participation of lipids in the transmembrane ion translocating process is lacking, and the small magnitude of the lipid effect is at variance with the kinetic parameters of the biological ion transport process. Apparently the 'ionophorous' property of these substances is just a 'side' effect.

Persistent efforts to isolate intrinsic ionophores from biological membranes are being made in recent years in D. Green's laboratory. In 1971, Blondin et al. [70] from this laboratory published a sensational paper on the extraction from bovine heart mitochondria, by organic solvents, of a Na^+, K^+-specific dodecapeptide with marked ionophorous activity. The absence of terminal groups suggested to the authors that this 'ionophore' should be quite analogous to valinomycin. However, complete identification and structural elucidation of the compound could not be achieved, owing to the negligible amounts of material they had available (~100–250 nanogram/gram bovine mitochondrial protein).

Later, Blondin [301] proposed that the Na^+, K^+ ionophore of mitochondria is a composite part of a complicated protein complex, which he termed 'ionophoroprotein'. The ionophore itself, i.e. the low molecular component of this complex, may be isolated in the free state, but only after considerable degradation of the ionophoroprotein with proteolytic enzyme; isolated in this way it displays ionophore activity in bilayers. This concept was extended also to other types of ionophores.

In 1974 Blondin [302] reported the isolation from mitochondria of a divalent cation in particular Ca^{2+} and Mg^{2+} ionophore (see also [303]). This lipid-soluble compound was obtained by tryptic digestion of the ionophore-active fraction of mitochondrial membranes. Subsequent careful purification and chemical study of the ionophore material thus isolated led to identification of a number of allied compounds clearly displaying the properties of classic ionophores (similar to X-537A and A-23187) and turning out to be derivatives of octadecadienic acid (Table 1).

The structures of these compounds are given in Fig. 20. In this figure the formulas of two prostaglandins namely PGB_1 and PGB_2 are also given for comparison. The close resemblance between these two types of compounds is striking, and the more so in light of the recent reports about the ionophorous activities of certain prostaglandins [304,305]. At the same time, however enticing it might be to consider that intrinsic ionophores, at least with respect to Ca^{2+} and Mg^{2+} ions have been discovered, it seems feasible to wait until the completion of this series of investigations before arriving at a final conclusion; particularly, as we have seen that the authors themselves have only partially identified these and the other, chemically different, ionophores which they had isolated from mitochondria, from the sarcoplasmatic reticulum and from chloroplasts [71,303].

If, on the other hand, the participation of prostaglandins in ion transport were confirmed, it would be still another factor adding to the interest in these bioregulators that have loomed in the horizon of biochemistry and have engendered many expectations in both theoretical and practical domains [306, 307].

Finally, recently it has been reported [73] that an electrogenic K^+/Ca^{2+} ionophore has been isolated from mitochondria and also on trypsinolysis of the corresponding ionophoroprotein. To all appearances this paper seems to be a

Table 1. Intrinsic Ca^{2+}, Mg^{2+} ionophores isolated from bovine heart mitochondria.

Structure	Yield (nmol per mg protein)
1. 9-Hydroxy-10-*trans*-12-*cis*-octadecadienic acid	1.4
2. 13-Hydroxy-9-*cis*-11-*trans*-octadecadienic acid	0.8
3. 9-Oxo-10,12-octadecadienic acid 13-Oxo-8,11-octadecadienic acid	0.67
4. 9-Hydroxy-10-*trans*-12-*cis*-octadienic acid methyl ester	0.28
5. 9-Oxo-10,12-octadienic acid methyl ester	0.12

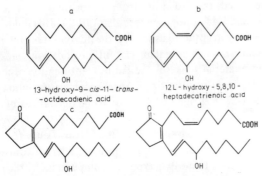

Fig. 20. The structure of (a and b) bovine heart mitochondrial ionophores and (c and d) prostaglandins B_1 and B_2.

modification of the previous reports [70,301] on this subject. The corresponding ionophoroprotein M.W. 10,000 was isolated by extracting lyophilized mitochondria with a 2 : 1 chloroform : methanol mixture. Its tryptic digestion yielded first a so-called ionophoropeptide (37 amino acid residues, M.W. 5,100), and then the ionophore itself (17 amino acid residues, M.W. 1,600). In its effect on mitochondrial K^+ ion transport, the isolated ionophore closely resembles valinomycin; however, it displays only K^+-ion specificity.

It should be stressed that, whereas this study [73] is more thorough than the previous investigations of these authors, it seems to smoothly pass over from attempts to isolate an intrinsic ionophore to, rather, isolation of a ion channel component; it, thereby, assumes a place in the ion channel searching field, rather than ionophore searching field [308–314].

Now, after such a tremendous effort to isolate an intrinsic ionophore has ended us up again at ion channels, may not the question be posed as to whether there exist in general intrinsic ionophores among the membrane proteins, and whether the search for such substances might not turn out fruitless. Here, too, it is difficult to arrive at an unequivocal answer, although a number of comparatively small proteins in various biological membranes deserve careful scrutiny.

Thus, some years ago, proteins, glycoproteins, to be more exact, were isolated from mitochondria, that were capable of selectively binding Ca^{2+} [315, 316]. It was suggested, and substantiated by a number of experiments, that these glycoproteins are active Ca^{2+} translocators. Yet all attempts to reassemble a Ca^{2+} transporting system in bilayers and liposomes have met with failure, and now these glycoproteins are regarded rather as specific surface Ca^{2+} receptors, and to see in this function their role in the Ca^{2+} translocating system in mitochondria [317–327].

Another type of possible candidates as ionophores are the so-called periplasmatic proteins of the bacterial cell. They are considered as participating in the specific active transport in bacteria of certain ions and low molecular weight metabolites, including amino acids and sugars, all with very high selectivity [328–333].

Such proteins have been isolated by cold osmotic shock and many have been purified to the individual state and are well characterized. Their molecular weight varies from 25,000 to 50,000. They are capable of binding their substrates in solution not only specifically, but very efficiently ($\sim 10^{-7}$ M). The binding mechanism, however, remains obscure. So far, a thorough study has been made of the proteins binding sulfate [334–336] and phosphate [337,338] ions, and over 15 amino acids [332,333], etc. Although their role in the transport process and the mechanism whereby they translocate the ions is still unknown, the concept of its being of a ionophorous nature is very attractive [339–344]. Certain grounds for such a belief might serve the fact of their compying with one of the most rigorous requirements for a ionophore: its ability to specifically bind the ion or substrate which it is transporting.

In a study of the structural principles underlying the functioning of amino acid binding proteins we selected two representatives of this class, the so-called LIV protein, a protein from E. coli binding leucine, isoleucine and valine, and a protein specifically binding leucine (LS protein).

First it was shown that these proteins are quite rigid in solution, retaining a given conformation even under considerable variation of the environmental conditions (pH, temperature, ion strength, etc.). Moreover, no significant conformational changes are observed in the protein on its binding leucine, thus providing strong evidence of the rigidity of the Leu-binding site [332] (Fig. 21).

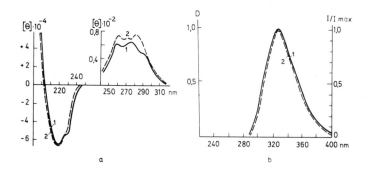

Fig. 21. The (a) CD and (b) fluorescence spectra of (1) LIV-protein and (2) its complex with leucine.

As a first step in the elucidation of the leucine-binding mechanism we carried out the complete structural determination of LIV protein (Fig. 22) [*345*], and a preliminary comparison of the amino acid sequences of the N-terminal and Cys-containing regions of LIV and LS proteins (Fig. 23). Unexpectedly, it turned out that the primary structures of the proteins in these two regions are very similar, although the proteins differ greatly in their valine and isoleucine binding capacity. One may expect that the extension of such a comparison will lay bare many more interesting facts about these proteins.

Not less informative should be the studies we have undertaken of the chemical modification of LIV protein. The first results have shown, for instance, the

$$\overset{20}{\text{Glu-Asp-Ile-Lys-Val-Ala-Val-Val-Gly-Ala-Met-Ser-Gly-Pro-Val-Ala-Gln-Tyr-Gly-Asp-}}$$

$$\overset{40}{\text{Gln-Glu-Phe-Thr-Gly-Ala-Glu-Gln-Ala-Val-Ala-Asp-Ile-Asn-Ala-Lys-Gly-Gly-Ile-Lys-}}$$

$$\overset{60}{\text{Gly-Asn-Lys-Leu-Gln-Ile-Val-Lys-Tyr-Asp-Asp-Ala-Cys-Asp-Pro-Lys-Gln-Ala-Val-Ala-}}$$

$$\overset{80}{\text{Val-Ala-Asn-Lys-Val-Val-Asn-Asp-Gly-Ile-Lys-Tyr-Val-Ile-Gly-His-Leu-Cys-Ser-Ser-}}$$

$$\overset{100}{\text{Ser-Thr-Gln-Pro-Ala-Ser-Asp-Ile-Tyr-Glu-Asp-Glu-Gly-Ile-Leu-Met-Ile-Thr-Pro-Ala-}}$$

$$\overset{120}{\text{Ala-Thr-Ala-Pro-Glu-Leu-Thr-Ala-Arg-Gly-Tyr-Gln-Leu-Ile-Leu-Arg-Thr-Thr-Gly-Leu-}}$$

$$\overset{140}{\text{Asp-Ser-Asp-Gln-Gly-Pro-Thr-Ala-Ala-Lys-Tyr-Ile-Leu-Glu-Lys-Val-Lys-Pro-Gln-Arg-}}$$

$$\overset{160}{\text{Ile-Ala-Ile-Val-His-Asp-Lys-Gln-Gln-Tyr-Gly-Glu-Gly-Leu-Ala-Arg-Ala-Val-Gln-Asp-}}$$

$$\overset{180}{\text{Gly-Leu-Lys-Lys-Gly-Asn-Ala-Asn-Val-Val-Phe-Phe-Asp-Gly-Ile-Thr-Ala-Gly-Glu-Lys-}}$$

$$\overset{200}{\text{Asp-Phe-Ser-Thr-Leu-Val-Ala-Arg-Leu-Lys-Lys-Glu-Asn-Ile-Asp-Phe-Val-Tyr-Tyr-Gly-}}$$

$$\overset{220}{\text{Gly-Tyr-His-Pro-Glu-Met-Gly-Gln-Ile-Leu-Arg-Gln-Ala-Arg-Ala-Ala-Gly-Leu-Lys-Thr-}}$$

$$\overset{240}{\text{Gln-Phe-Met-Gly-Pro-Glu-Gly-Val-Ala-Asn-Val-Ser-Leu-Ser-Asn-Ile-Ala-Gly-Glu-Ser-}}$$

$$\overset{260}{\text{Ala-Glu-Gly-Leu-Leu-Val-Thr-Lys-Pro-Lys-Asn-Tyr-Asp-Gln-Val-Pro-Ala-Asn-Lys-Pro-}}$$

$$\overset{280}{\text{Ile-Val-Ala-Asp-Ile-Lys-Ala-Lys-Lys-Gln-Asp-Pro-Ser-Gly-Ala-Phe-Val-Trp-Thr-Thr-}}$$

$$\overset{300}{\text{Tyr-Ala-Ala-Leu-Gln-Ser-Leu-Gln-Ala-Gly-Leu-Asn-Gln-Ser-Asp-Asp-Pro-Ala-Glu-Ile-}}$$

$$\overset{320}{\text{Ala-Lys-Tyr-Leu-Lys-Ala-Asn-Ser-Val-Asp-Thr-Val-Met-Gly-Pro-Leu-Thr-Trp-Asp-Glu-}}$$

$$\overset{340}{\text{Lys-Gly-Asp-Leu-Lys-Gly-Phe-Glu-Phe-Gly-Val-Phe-Asp-Trp-His-Ala-Asn-Gly-Thr-Ala-}}$$

Thr-Ala-Asp-Lys

Fig. 22. The primary structure of LIV-protein.

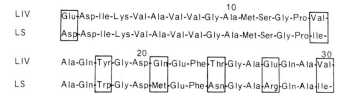

Fig. 23. The N-terminal sequence of LIV- and LS-proteins.

important role of lysine and tyrosine residues [346] in the binding of the substrate.

One can, thus, expect that the detailed study, first of the primary and then of the spatial structure of the binding proteins, will provide the key to the mode of action of these systems in biological membranes (cf. [347,348]).

In summary it may be said that the discovery of ionophores has led to a new, highly illuminating direction in physicochemical biology. The depth of our understanding of the underlying principles in this area has not been paralleled in most other fields. Neither are there many substances in biology that have so fruitfully lent themselves to study of the structure-function correlation or to such precise planning and such beneficial use of the composite approach to uncover their mode of action. It has thus turned out that the chemical and biochemical studies of the ionophores, these elegant molecules with such diverse properties, have not only deepened our understanding in general of the structure-function relationship in bioorganic chemistry, but have been the determining factor in a whole period of study of the theoretical and general biological basis of transmembrane ion transport. They have also served as the centripetal force about which has developed an extensive school of investigators, still continuing work on urgent problems of membrane transport. Whatever turn will take the fate of this extensive class of substances in the future, whether it will turn out that Nature has indeed made use of a mechanism resembling theirs for ion transport in biological membranes or is utilizing some other of their unique properties, it may be said with full justification that modern membranology will yet more than once harvest the fruit of this richly yielding field that has been so abundantly shown in the spring of ionophores study.

Acknowledgement

The author is grateful to Dr. G. Peck for his critical translation of the Russian text into English.

References

1. Ovchinnikov Yu.A., Ivanov V.T., Shkrob A.M. (1974) *Membrane Active Complexones*, BBA Library, Vol. 12. Elsevier, Amsterdam.
2. Danielli J.F., Davson H. (1935) J. Cell. Comp. Physiol., *5*, 495.
3. Davson H., Danielli J.F. (1952) *The Permeability of Natural Membranes*. Cambridge University Press, London.
4. Robertson I.D. (1964) In: M. Locke, ed.: *Cellular Membranes in Development*. Academic Press, New York, N.Y., p. 1.
5. Wallach D.F.H., Zahler P.H. (1966) Proc. Natl. Acad. Sci. USA *56*, 1552.
6. Lenard J., Singer S.J. (1966) Proc. Natl. Acad. Sci. USA *56*, 1828.
7. Singer S.J. (1971) In: L.I. Rothfield, ed.: *Structure and Function of Biological Membranes*. Academic Press, New York, N.Y., p. 145.
8. Singer S.J., Nicolson G.L. (1972) Science *175*, 720.
9. Singer S.J. (1977) In: S. Abrahamson, I. Pascher, eds.: *Structure of Biological Membranes*. Plenum Press, New York, N.Y., p. 443.
10. Ansell G.B., Hawthorne J.N., Dawson R.M.C. (1973) *Form and Function of Phospholipids*. Elsevier, Amsterdam.
11. Kuksis A. (1977) J. Chromatogr. *143*, 3.
12. G.V. Marinetti, ed.: *Lipid Chromatographic Analysis*, Vol. 1. M. Dekker Inc., New York, N.Y., 1976.
13. Guidotti G. (1972) Ann. Rev. Biochem. *41*, 731.
14. Tanford C. (1977) In: S. Abrahamson, I. Pascher, eds.: *Structure of Biological Membranes*. Plenum Press, New York, N.Y., p. 497.
15. Wallach D.F.H., Bieri V., Verma S.P., Schmidt-Ullrich R. (1975) Ann. N.Y. Acad. Sci. *264*, 142.
16. Kornberg R.D., McConnell H.M. (1971) Biochemistry *10*, 1111.
17. Overath P., Träuble H. (1973) Biochemistry *12*, 2625.
18. Ehrenberg A., Shimoyama Y., Eriksson L.E.G. (1977) In: S. Abrahamson, I. Pascher, eds.: *Structure of Biological Membranes*. Plenum Press, New York, N.Y., p. 119.
19. Hackenbrock C.R., Höchli M. (1977) In: G. Semenza, E. Carafoli, eds.: *Biochemistry of Membrane Transport*. Springer Verlag, Berlin, Heidelberg, New York, p. 10.
20. Chapman D., Dodd G.H. (1971) In: L.I. Rothfield, ed.: *Structure and Function of Biological Membranes*. Academic Press, New York, N.Y., p. 13.
21. Lee A.G. (1975) In: J.A. Butler, D. Noble, eds.: *Progress in Biophysics and Molecular Biology*. Vol. 29, Pergamon Press, New York, N.Y., p. 3.
22. Thompson T.E., Lentz B.R., Barenholz Y. (1977) In: G. Semenza, E. Carafoli, eds.: *Biochemistry of Membrane Transport*. Springer-Verlag, Berlin, Heidelberg, New York, p. 47.
23. Mitchell P. (1966) Biol. Rev. *41*, 445.
24. Mitchell P. (1973) FEBS Lett. *33*, 267.
25. Mitchell P. (1975) FEBS Lett. *50*, 95.
26. Mitchell P. (1976) Biochem. Soc. Trans. *4*, 399.
27. Avron M. (1976) In: Y. Hatefi, L. Djavadi-Ohaniance, eds.: *Structural Basis of Membrane Function*. Academic Press, New York, N.Y., p. 227.

28. Ernster L., Asami K., Juntti K., Coleman J., Nordenbrand K. (1977) In: S. Abrahamson, I. Pascher, eds.: *Structure of Biological Membranes*. Plenum Press, New York, N.Y., p. 135.
29. Drachev L.A., Kondrashin A.A., Samuilov V.D., Skulachev V.P. (1975) FEBS Lett. *50*, 219.
30. Skulachev V.P. (1978) In: D.C. Tosteson, Yu.A. Ovchinnikov, R. Latorre, eds.: *Membrane Transport Processes*, Vol. 2. Raven Press, New York, N.Y., p. 385.
31. Eytan G.D., Schatz G., Racker E. (1977) In: S. Abrahamson, I. Pascher, eds.: *Structure of Biological Membranes*. Plenum Press, New York, N.Y., p. 373.
32. Racker E., Eytan E. (1975) J. Biol. Chem. *250*, 7533.
33. Yoshida M., Okamoto H., Sone N., Hirata H., Kagawa Y. (1977) Proc. Natl. Acad. Sci. USA *74*, 936.
34. Sone N., Yoshida M., Hirata H., Kagawa Y. (1977) J. Biol. Chem. *252*, 2956.
35. Steck T.L. (1978) In: D.C. Tosteson, Yu.A. Ovchinnikov, R. Latorre, eds.: *Membrane Transport Processes*, Vol. 2, Raven Press, New York, N.Y., p. 39.
36. Kaback H.R. (1970) Ann. Rev. Biochem. *39*, 561.
37. Kaback H.R. (1974) Science *186*, 882.
38. Kaback H.R. (1977) In: G. Semenza, E. Carafoli, eds.: *Biochemistry of Membrane Transport*. Springer Verlag, Berlin, Heidelberg, New York, p. 598.
39. Mueller P., Rudin D.O., Tien H.T., Wescott W.C. (1962) Nature (London) *194*, 979.
40. Mueller P., Rudin D.O., Tien H.T., Wescott W.C. (1963) J. Phys. Chem. *67*, 534.
41. Mueller P., Rudin D.O. (1963) J. Theoret. Biol. *4*, 268.
42. Bangham A.D., Standish M.M., Watkins J.C. (1965) J. Mol. Biol. *13*, 238.
43. Bangham A.D. (1968) In: J.A. Butler, D. Noble, eds.: *Progress in Biophysics and Molecular Biology*, Vol. 18. Pergamon Press, New York, N.Y., p. 29.
44. Pressman B.C. (1973) Fed. Proc. *32*, 1698.
45. Moore C., Pressman B.C. (1964) Biochem. Biophys. Res. Commun. *15*, 562.
46. Pressman B.C. (1965) Fed. Proc. *24*, 425.
47. Pressman B.C. (1965) Proc. Natl. Acad. Sci. USA *53*, 1076.
48. Pressman B.C., Harris E.J., Jagger W.S., Johnson J.H. (1967) Proc. Natl. Acad. Sci. USA *58*, 1949.
49. Pressman B.C., deGuzman N.T. (1975) Ann. N.Y. Acad. Sci. *264*, 373.
50. Akasaki K., Karasawa K., Watanabe M., Yonehara H., Umezawa H. (1963) J. Antibiot. Ser. A *16*, 127.
51. Lardy H.A., Graven S.N., Estrada-O S. (1968) Fed. Proc. *26*, 1355.
52. Pressman B.C. (1968) Fed. Proc. *27*, 1283.
53. McLaughlin S., Eisenberg M. (1975) Ann. Rev. Biophys. Bioeng. *4*, 335.
54. Hladky S.B., Haydon D.A. (1972) Biochim. Biophys. Acta *274*, 294.
55. Eisenman G., Sandblom J., Neher E. (1978) In: D.C. Tosteson, Yu.A. Ovchinnikov, R. Latorre, eds.: *Membrane Transport Processes*, Vol. 2. Raven Press, New York, N.Y., p. 285.

56. Mueller P. (1975) Ann. N.Y. Acad. Sci. *264*, 247.
57. Finkelstein A., Holz R. (1973) In: G. Eisenman, ed.: *Membranes – A Series of Advances*, Vol. 2. Marcel Dekker, New York, N.Y., p. 377.
58. Mueller P., Rudin D.O. (1967) Biochem. Biophys. Res. Commun. *26*, 398.
59. Lev A.A., Buzhinsky E.P. (1967) Cytologia (USSR) *9*, 102.
60. Eisenman G., Szabo G., Ciani S., McLaughlin S., Krasne S. (1973) In: J.F. Danielli, M.D. Rosenberg, D.A. Cadenhead, eds.: *Progr. Surf. Membr. Sci.* Academic Press, New York, N.Y., p. 139.
61. Ciani S.M., Eisenman G., Laprade R., Szabo G. (1973) In: G. Eisenman, ed.: *Membranes – A Series of Advances*, Vol. 2. Marcel Dekker, New York, N.Y., p. 61.
62. Benz R., Stark G., Ganko K., Läuger P. (1973) J. Membr. Biol. *14*, 339.
63. Eisenman G., Ciani S.M., Szabo G. (1968) Fed. Proc. *27*, 1289.
64. Eisenman G., Szabo G., McLaughlin S.G.A., Ciani S.M. (1972) Bioenergetics Bull. *4*, 295.
65. Diebler H., Eigen M., Ilgenbretz G., Maas G., Winkler R. (1969) Pure Appl. Chem. *20*, 93.
66. Pressman B.C. (1976) Ann. Rev. Biochem. *45*, 501.
67. Grell E., Oberbäumer I., Ruf H., Zingsheim H.P. (1977) In: G. Semenza, E. Carafoli, eds.: *Biochemistry of Membrane Transport*. Springer Verlag, Berlin, Heidelberg, New York, p. 147.
68. Benz R., Stark G. (1975) Biochim. Biophys. Acta *382*, 27.
69. Benz R., Gisin B.F., Ting-Beall H.P., Tosteson D.C., Läuger P. (1976) Biochim. Biophys. Acta *455*, 665.
70. Blondin G.A., DeCastro A.F., Senior A.E. (1971) Biochem. Biophys. Res. Commun. *43*, 28.
71. Green D.E. (1975) Ann. N.Y. Acad. Sci. *264*, 61.
72. Green D.E., Blondin G., Kessler R., Southard J.H. (1975) Proc. Natl. Acad. Sci. USA *72*, 896.
73. Blondin G.A., Kessler R.J., Green D.E. (1977) Proc. Natl. Acad. Sci. USA *74*, 3667.
74. Osterhout W.J.V., Stanley W.M. (1932) J. Gen. Physiol. *15*, 667.
75. Osterhout W.J.V. (1935) Proc. Natl. Acad. Sci. USA *21*, 125.
76. Shemyakin M.M., Ovchinnikov Yu.A., Ivanov V.T., Antonov V.K., Vinogradova E.I., Shkrob A.M., Malenkov G.G., Evstratov A.V., Ryabova I.D., Laine I.A., Melnik E.I. (1969) J. Membr. Biol. *1*, 402.
77. Ivanov V.T. (1975) Ann. N.Y. Acad. Sci. *264*, 221.
78. Weinbach E.C. (1954) J. Biol. Chem. *210*, 545.
79. Heytler P.G. (1963) Biochemistry *2*, 357.
80. Bielawski J., Thompson T.E., Lehninger A.L. (1966) Biochem. Biophys. Res. Commun. *24*, 948.
81. Woodruff R.C.S., Wilkinson B.E. (1966) J. Gen. Microbiol. *44*, 343.
82. Beechey R.B. (1966) Biochem. J. *98*, 284.
83. Skulachev V.P., Sharaf A.A., Liberman E.A. (1967) Nature (London) *216*, 718.
84. Harold F.M., Baarda J.R. (1968) J. Bacteriol. *95*, 816.
85. Liberman E.A., Topaly V.P. (1968) Biofizika (USSR) *13*, 1025.

86. Hopfer U., Lehninger A.L., Thompson T.E. (1968) Proc. Natl. Acad. Sci. USA *59*, 484.
87. Liberman E.A., Topaly V.P., Silberstein A.Y. (1970) Biochim. Biophys. Acta *196*, 221.
88. Liberman E.A., Topaly V.P., Silberstein A.Y., Okhlobystin O.Yu. (1971) Biofizika (USSR) *16*, 615.
89. Yaguzhinsky L.S., Boguslavsky L.I., Ismailov A.D. (1974) Biochim. Biophys. Acta *368*, 22.
90. Anderson S.S., Lyle I.G., Paterson R. (1976) Nature (London) *259*, 147.
91. Hauska G., Trebst A. (1977) In: D.R. Sanadi, ed.: *Current Topics in Bioenergetics*, Vol. 6. Academic Press, New York, N.Y., p. 151.
92. Hauska G. (1977) FEBS Lett. *79*, 345.
93. Mitchell P. (1977) FEBS Lett. *78*, 1.
94. Shemyakin M.M., Ovchinnikov Yu.A., Ivanov V.T., Evstratov A.V. (1967) Nature (London) *213*, 413.
95. Ivanov V.T., Laine I.A., Ryabova I.D., Ovchinnikov Yu.A. (1970) Khim. Prir. Soedin. (USSR) 744.
96. Prelog V. (1977) *Plenary Lecture to the 50th-Anniversary Meeting of the Serbian Chemical Society*, Beograd.
97. Brockmann H., Schmidt-Kastner G. (1955) Chem. Ber. *88*, 57.
98. Shemyakin M.M., Aldanova N.A., Vinogradova E.I., Feigina M.Yu. (1963) Tetrahedron Lett. 1921.
99. Shemyakin M.M., Ovchinnikov Yu.A., Ivanov V.T., Kiryushkin A.A., Zhdanov G.L., Ryabova I.D. (1963) Experientia *19*, 566.
100. Shemyakin M.M., Vinogradova E.I., Feigina M.Yu., Aldanova N.A. (1964) Zh. Obshch. Khim. (USSR) *34*, 1798.
101. Shemyakin M.M., Vinogradova E.I., Feigina M.Yu., Aldanova N.A., Shvetsov Yu.B., Fonina L.A. (1966) Zh. Obshch. Khim. (USSR) *36*, 1391.
102. McMurray W., Begg R.W. (1959) Arch. Biochem. Biophys. *84*, 546.
103. Ivanov V.T., Laine I.A., Abdullaev N.D., Senyavina L.B., Popov E.M., Ovchinnikov Yu.A., Shemyakin M.M. (1969) Biochem. Biophys. Res. Commun. *34*, 803.
104. Pinkerton M., Steinrauf L.K., Dawkins P. (1969) Biochem. Biophys. Res. Commun. *35*, 512.
105. Ovchinnikov Yu.A., Ivanov V.T. (1975) Tetrahedron *31*, 2177.
106. Ohnishi M., Urry D.W. (1969) Biochem. Biophys. Res. Commun. *36*, 194.
107. Ohnishi M., Urry D.W. (1970) Science *168*, 1091.
108. Patel D.J., Tonnelli A. (1973) Biochemistry *12*, 486.
109. Patel D.J. (1973) Biochemistry *12*, 496.
110. Grell E., Funck Th. (1973) J. Supramol. Struct. *1*, 307.
111. Rothschild K.J., Asher I.M., Anastassakis E., Stanley H.E. (1973) Science *182*, 384.
112. Asher I.M., Rothschield K.J., Stanley H.E. (1974) J. Mol. Biol. *89*, 205.
113. Asher I.M., Rothschield K.J., Anastassakis E., Stanley H.E. (1977) J. Amer. Chem. Soc. *99*, 2024.
114. Rothschield K.J., Asher I.M., Stanley H.E., Anastassakis E. (1977) J. Amer. Chem. Soc. *99*, 2032.

115. Duax W.L., Hauptman H., Weeks C.M., Norton D.A. (1972) Science *176*, 911.
116. Smith G.D., Duax W.L., Langs D.A., DeTitta G.T., Edmonds J.W., Rohrer D.C., Weeks C.M. (1975) J. Amer. Chem. Soc. *97*, 7242.
117. Karle I.L. (1975) J. Amer. Chem. Soc. *97*, 4379.
118. Neupert-Laves K., Dobler M. (1975) Helv. Chim. Acta *58*, 432.
119. Pletnev V.Z., Galitsky N.M., Ivanov V.T., Ovchinnikov Yu.A. (1977) Bioorgan. Khim. (USSR) *3*, 1427.
120. Ovchinnikov Yu.A. (1971) In: *XXIIIrd International Congress of Pure and Applied Chemistry*, Vol. 2, Butterworths, London, p. 121.
121. Ovchinnikov Yu.A. (1974) FEBS Lett. *44*, 1.
122. Ovchinnikov Yu.A. (1977) In: S. Abrahamsson, I. Pascher, eds.: *Structure of Biological Membranes*. Plenum Press, New York, N.Y., p. 345.
123. Ovchinnikov Yu.A., Ivanov V.T. (1977) In: G. Semenza, E. Carafoli, eds.: *Biochemistry of Membrane Transport*. Springer Verlag, Berlin, Heidelberg, New York, p. 123.
124. Gisin B.F., Merrifield R.B. (1972) J. Amer. Chem. Soc. *94*, 6165.
125. Davis D.G., Gisin B.F., Tosteson D.C. (1976) Biochemistry *15*, 768.
126. Ivanov V.T., Sanasaryan A.A., Chervik I.M., Yakovlev G.I., Fonina L.A., Senyavina L.N., Sychev S.V., Vinogradova E.I., Ovchinnikov Yu.A. (1974) Izv. Akad. Nauk SSSR, ser. Khim. (USSR) 2310.
127. Vinogradova E.I., Fonina L.A., Ryabova I.D., Ivanov V.T. (1974) Khim. Prir. Soedin. (USSR) 278.
128. Melnik E., Latorre R., Hall J.E., Tosteson D.C. (1977) J. Gen. Physiol. *69*, 243.
129. Fonina L.A., Savelov I.S., Avotina G.Ya., Ivanov V.T., Ovchinnikov Yu.A. (1976) In: A. Loffet, ed.: *Peptides 1976*. Ed. de l'Université de Bruxelles, p. 635.
130. Eisenman G., Krasne S., Ciani S. (1976) In: M. Kessler, L. Clark, D. Lübbers, I. Silver, W. Simon, eds.: *Ion Selective Electrodes and Enzyme Electrodes in Medicine and in Biology*. Urban and Schwarzenberg, Munich, Vienna, p. 3.
131. Plattner Pl.A., Nager, U. (1947) Experientia *3*, 325.
132. Cook A.H., Cox S.F., Farmer T.H., Lacey M.S. (1947) Nature (London) *160*, 31.
133. Shemyakin M.M., Ovchinnikov Yu.A., Ivanov V.T., Kiryushkin A.A. (1963) Tetrahedron Lett. 885.
134. Shemyakin M.M., Ovchinnikov Yu.A., Kiryushkin A.A., Ivanov V.T. (1963) Izv. Akad. Nauk SSSR, ser. Khim. (USSR) 579.
135. Shemyakin M.M., Ovchinnikov Yu.A., Kiryushkin A.A., Ivanov V.T. (1963) Izv. Akad. Nauk SSSR, ser. Khim. (USSR) 1148.
136. Dobler M., Dunitz J.D., Krajewski J. (1969) J. Mol. Biol. *42*, 603.
137. Ovchinnikov Yu.A., Ivanov V.T., Evstratov A.V., Bystrov V.F., Abdullaev N.D., Popov E.M., Lipkind G.M., Arkhipova S.F., Efremov E.S., Shemyakin M.M. (1969) Biochem. Biophys. Res. Commun. *37*, 668.
138. Ovchinnikov Yu.A., Ivanov V.T., Antonov V.K., Shkrob A.M., Mikhaleva I.I., Evstratov A.V., Malenkov G.G., Melnik E.I., Shemyakin M.M. (1968)

In: E. Bricas, ed.: *Peptides, Proc. IXth European Peptide Symposium*. North-Holland Publ. Co., Amsterdam, p. 56.
139. Grell E., Funk Th., Eggers F. (1972) In: E. Muñoz, F. Garcia-Ferrándiz, D. Vazquez, eds.: *Molecular Mechanisms of Antibiotic Action on Protein Biosynthesis and Membranes*. Elsevier, Amsterdam, p. 646.
140. Ivanov V.T., Evstratov A.V., Sumskaya L.V., Melnik E.I., Chumburidze T.S., Portnova S.L., Balashova T.A., Ovchinnikov Yu.A. (1973) FEBS Lett. *36*, 65.
141. Ovchinnikov Yu.A., Ivanov V.T., Evstratov A.V., Mikhaleva I.I., Bystrov V.F., Portnova S.L., Balashova T.A., Meshcheryakova E.A., Tulchinsky V.M. (1974) Int. J. Pept. Protein Res. *6*, 465.
142. Sumskaya L.V., Balashova T.A., Mikhaleva I.I., Chumburidze T.S., Melnik E.I., Ivanov V.T., Ovchinnikov Yu.A. (1977) Bioorgan. Khim. (USSR) *3*, 5.
143. Tishchenko G.N., Karimov Z., Vainshtein B.K., Evstratov A.V., Ivanov V.T., Ovchinnikov Yu.A. (1976) FEBS Lett. *65*, 315.
144. Shishova T.G., Simonov V.I., Ivanov V.T., Evstratov A.V., Mikhaleva I.I., Balashova T.A., Ovchinnikov Yu.A. (1975) Bioorgan. Khim. (USSR) *1*, 1689.
145. Wieland Th., Lüben G., Ottenheym H., Faesel J., De Vries J.X., Konz W., Prox A., Schmid J. (1968) Angew. Chem. *80*, 209.
146. Wieland Th., Faulstich H., Burgermeister W., Otting W., Möhle W., Shemyakin M.M., Ovchinnikov Yu.A., Ivanov V.T., Malenkov G.G. (1970) FEBS Lett. *9*, 89.
147. Wieland Th. (1972) Naturwissenschaften *59*, 225.
148. Wieland Th., Govindan V.M. (1974) FEBS Lett. *46*, 513.
149. Faulstich H., Wieland Th., Walli A., Birkmann K. (1974) Hoppe-Seyler's Z. Physiol. Chem. *355*, 1162.
150. Pressman B.C. (1973) In: G.L. Eichhorn, ed.: *Inorganic Biochemistry*. Elsevier, Amsterdam, p. 203.
151. Ivanov V.T., Miroshnikov A.I., Abdullaev N.D., Senyavina L.B., Arkhipova S.F., Uvarova N.N., Khalilulina K.Kh., Bystrov V.F., Ovchinnikov Yu.A. (1971) Biochem. Biophys. Res. Commun. *42*, 654.
152. Ivanov V.T., Miroshnikov A.I., Kozmin S.A., Meshcheryakova E.A., Senyavina L.B., Uvarova N.N., Khalilulina K.Kh., Zabrodin V.A., Bystrov V.F., Ovchinnikov Yu.A. (1973) Khim. Prir. Soedin. (USSR) 378.
153. Patel D.J. (1973) Biochemistry *12*, 677.
154. Patel D.J., Tonelli A.E. (1974) Biochemistry *13*, 788.
155. Karle I.L., Karle J., Wieland Th., Burgermeister W., Faulstich H., Witkop B. (1973) Proc. Natl. Acad. Sci. USA *70*, 1836.
156. Karle I.L. (1974) J. Amer. Chem. Soc. *96*, 4000.
157. Karle I.L. (1974) Biochemistry *13*, 2155.
158. Karle I.L. (1978) In: D.C. Tosteson, Yu.A. Ovchinnikov, R. Latorre, eds.: *Membrane Transport Processes*, Vol. 2. Raven Press, New York, N.Y., p. 247.
159. Karle I.L., Duesler E. (1977) Proc. Natl. Acad. Sci. USA *74*, 2602.
160. Karle I.L. (1977) J. Amer. Chem. Soc. *99*, 5152.
161. Ovchinnikov Yu.A., Ivanov V.T., Barsukov L.I., Melnik E.I., Oreshnikova

N.I., Bogolyubova N.D., Ryabova I.D., Miroshnikov A.I., Rimskaya V.A. (1972) Experientia *28*, 399.
162. Deber C.M., Torchia D.A., Wong S.C.K., Blout E.R. (1972) Proc. Natl. Acad. Sci. USA *69*, 1825.
163. Ivanov V.T., Lavrinovich I.A., Portnova S.L., Sychev S.V., Lapshin V.V., Kostetskii P.V., Ovchinnikov Yu.A. (1975) Bioorgan. Khim. (USSR) *1*, 149.
164. Hallosi M., Wieland Th. (1977) Int. J. Pept. Protein Res. *10*, 329.
165. Madison V., Deber C.M., Blout E.R. (1977) J. Amer. Chem. Soc. *99*, 4788.
166. Madison V., Atreyi M., Deber C.M., Blout E.R. (1974) J. Amer. Chem. Soc. *96*, 6725.
167. Baron D., Pease L.G., Blout E.R. (1977) J. Amer. Chem. Soc. *99*, 8299.
168. Schwyzer R., Tun-Kyi A., Caviezel M., Moser P. (1970) Helv. Chim. Acta *53*, 15.
169. Ammann D., Bissig R., Cimerman Z., Fiedler U., Güggi M., Morf W.E., Oehme M., Osswald H., Pretsch Z., Simon W. (1976) In: M. Kessler et al., eds.: *Ion and Enzyme Electrodes in Biology and Medicine*. Univ. Park Press, Baltimore, London, Tokyo, p. 22.
170. Vuilleumier P., Gazzotti P., Carafoli E., Simon W. (1977) Biochim. Biophys. Acta *467*, 12.
171. Caroni P., Gazzotti P., Vuilleumier P., Simon W., Carafoli E. (1977) Biochim. Biophys. Acta *470*, 437.
172. Borowitz I.J., Lin W.-C., Wun T.-C., Bittman R., Weiss L., Diakiw V., Borowitz G.B. (1977) Tetrahedron *33*, 1697.
173. Wun T.-C., Bittman R., Borowitz I.J. (1977) Biochemistry *16*, 2074.
174. Wun T.-C., Bittman R. (1977) Biochemistry *16*, 2080.
175. Kirsch N.N.L., Funck R.J.J., Pretsch E., Simon W. (1977) Helv. Chim. Acta *60*, 2326.
176. Corbaz R., Ettlinger L., Gäumann E., Keller-Schierlein W., Kradolfer F., Neipp L., Prelog V., Zähner H. (1955) Helv. Chim. Acta *38*, 1445.
177. Dominguez J., Dunitz J.D., Gerlach H., Prelog V. (1962) Helv. Chim. Acta *45*, 129.
178. Beck J., Gerlach H., Prelog V., Voser W. (1962) Helv. Chim. Acta *45*, 620.
179. Gerlach H., Prelog V. (1963) Justus Liebigs Ann. Chem. *669*, 121.
180. Gerlach H., Huber E. (1967) Helv. Chim. Acta *50*, 2087.
181. Kilbourn B.T., Dunitz J.D., Pioda L.A.R., Simon W. (1967) J. Mol. Biol. *30*, 559.
182. Dobler M., Dunitz J.D., Kilbourn B.T. (1969) Helv. Chim. Acta *52*, 2573.
183. Pedersen C.J. (1967) J. Amer. Chem. Soc. *89*, 2495.
184. Pedersen C.J. (1967) J. Amer. Chem. Soc. *89*, 7017.
185. Pedersen C.J. (1970) J. Amer. Chem. Soc. *92*, 386.
186. Sugiura M., Shinbo T. (1976) No Kagaku (Agricultural Chem.) *50*, 547.
187. Harris E.J., Zaba B., Truter M.R., Parsons D.G., Wingfield J.N. (1977) Arch. Biochem. Biophys. *182*, 311.
188. Frensdorff H.K. (1971) J. Amer. Chem. Soc. *93*, 600.
189. Frensdorff H.K. (1971) J. Amer. Chem. Soc. *93*, 4684.
190. Christensen J.J., Hill J.O., Izatt R.M. (1971) Science *174*, 459.

191. Izatt R.M., Terry R.E., Haymore B.L., Hansen L.D., Dalley N.K., Avondet A.G., Christensen J.J. (1976) J. Amer. Chem. Soc. *98*, 7620.
192. Izatt R.M., Terry R.E., Nelson D.P., Chan Y., Eatough D.J., Bradshaw J.S., Hansen L.D., Christensen J.J. (1976) J. Amer. Chem. Soc. *98*, 7626.
193. Gokel G.W., Durst H.D. (1976) Synthesis 168.
194. Gokel G.W., Durst H.D. (1976) Aldrichimica Acta *9*, 3.
195. Gokel G.W., Garcia B.J. (1977) Tetrahedron Lett. 317.
196. Gokel G.W., Korzeniowski S.H., Blum L. (1977) Tetrahedron Lett. 1633.
197. Korzeniowski S.H., Gokel G.W. (1977) Tetrahedron Lett. 3519.
198. Hayward R.J., Htay M.M., Meth-Cohn O. (1977) Chem. & Ind. (London) 373.
199. Vögtle F., Sieger H. (1977) Angew. Chem. *89*, 410.
200. Frensch K., Vögtle F. (1977) Tetrahedron Lett. 2573.
201. Reinhoudt D.N., Gray R.T., DeJong F., Smit C.J. (1977) Tetrahedron *33*, 563.
202. Cram D.J., Cram J.M. (1974) Science *183*, 801.
203. Madan K., Cram D.J. (1975) J. Chem. Soc. D, Chem. Commun. 427.
204. Kyba E.B., Koga K., Sousa L.R., Siegel M.G., Cram D.J. (1973) J. Amer. Chem. Soc. *95*, 2692.
205. Helgeson R.C., Koga K., Timko J.M., Cram D.J. (1973) J. Amer. Chem. Soc. *95*, 3021.
206. Newcomb M., Helgeson R.C., Cram D.J. (1974) J. Amer. Chem. Soc. *96*, 7367.
207. Cram D.J., Helgeson R.C., Sousa L.R., Timko J.M., Newcomb M., Moreau P., De Jong F., Gokel G.W., Hoffman D.H., Domeier L.A., Peacock S.C., Madan K., Kaplan L. (1975) Pure Appl. Chem. *43*, 327.
208. Timko J.M., Moore S.S., Walba D.M., Hiberty P.C., Cram D.J. (1977) J. Amer. Chem. Soc. *99*, 4207.
209. Thoma A.P., Cimerman Z., Fiedler Y., Bedekovič D., Güggi M., Jordan P., May K., Pretsch E., Prelog V., Simon W. (1975) Chimia *29*, 344.
210. Thoma A.P., Pretsch E., Horvai G., Simon W. (1977) In: G. Semenza, E. Carafoli, eds.: *Biochemistry of Membrane Transport*. Springer Verlag, Berlin, Heidelberg, New York, p. 116.
211. Pioda L.A.R., Stankova V., Simon W. (1969) Anal. Lett. *2*, 665.
212. Scholer R.P., Simon W. (1970) Chimia *24*, 372.
213. Ammann D., Pretsch E., Simon W. (1972) Tetrahedron Lett. 2473.
214. Wong L., Yagi K., Smid J. (1974) J. Membr. Biol. *18*, 379.
215. Burden I.J., Coxon A.C., Stoddart J.F., Wheatley C.M. (1977) J. Chem. Soc. (Perkin Trans. I) 220.
216. Curtis W.D., Laidler D.A., Stoddart J.F., Jones G.H. (1977) J. Chem. Soc. (Perkin Trans. I) 1756.
217. Wong L., Smid J. (1977) J. Amer. Chem. Soc. *99*, 5637.
218. Van Bergen T.J., Kellogg R.M. (1977) J. Amer. Chem. Soc. *99*, 3882.
219. Idemori K., Takagi M., Matsuda T. (1977) Bull. Chem. Soc. Japan *50*, 1355.
220. McKervey M.A., Mulholland D.L. (1977) J. Chem. Soc. D, Chem. Commun. 438.
221. Pitha J., Smid J. (1976) Biochim. Biophys. Acta *425*, 287.

222. Chang J., Kiesel R.F., Hogen-Esch T.E. (1975) J. Amer. Chem. Soc. 97, 2805.
223. Clement D., Damm F., Lehn J.-M. (1976) Heterocycles 5, 477.
224. Negishi E., Baba S. (1976) J. Chem. Soc. D, Chem. Commun. 596.
225. Cinouini M., Colonna S., Molinari H., Montanari F. (1976) J. Chem. Soc. D, Chem. Commun. 394.
226. Korzeniowski S.H., Blum L., Gokel G.W. (1977) Tetrahedron Lett. 1871.
227. Lehn J.-M., Sauvage J.P. (1971) J. Chem. Soc. D, Chem. Commun. 440.
228. Dietrich B., Lehn J.-M. (1973) Tetrahedron Lett. 1225.
229. Dietrich B., Lehn J.-M., Sauvage J.P., Blanzat J. (1973) Tetrahedron 29, 1629.
230. Dietrich B., Lehn J.-M., Sauvage J.P. (1973) Tetrahedron 29, 1647.
231. Graf E., Lehn J.-M. (1975) J. Amer. Chem. Soc. 97, 5022.
232. Kappenstein C. (1974) Bull. Soc. Chim. France 89.
233. Lehn J.-M., Simon J. (1977) Helv. Chim. Acta 60, 141.
234. Neilands J.B. (1973) In: G.L. Eichhorn, ed.: Inorganic Biochemistry, Vol. 1. Elsevier, Amsterdam, Chapt. 5.
235. Harned R.L., Hidy P.H., Corum C.J., Jones K.L. (1950) Proc. Indiana Acad. Sci. 59, 38.
236. Harned R.L., Hidy P.H., Corum C.J., Jones K.L. (1951) Antibiot. Chemother. 1, 594.
237. Graven S.N., Estrada-O S., Lardy H.A. (1966) Proc. Natl. Acad. Sci. USA 56, 654.
238. Seto H., Yahagi T., Miyazaki Y., Otaka N. (1977) J. Antibiot. 30, 530.
239. Koyama H., Utsumi-Oda K. (1977) J. Chem. Soc. 1531.
240. Otake N., Koenuma M., Miyamae H., Sato S., Saito Y. (1977) J. Chem. Soc. 494.
241. Otake N., Nakayama H., Miyamae H., Sato S., Saito Y., (1977) Chem. Commun. 590.
242. Zieniawa T., Popinigis J., Wozniak M., Cybulska B., Borowski E. (1977) FEBS Lett. 76, 81.
243. Mitani M., Yamanishi T., Ebata E., Otake N., Koenuma M. (1977) J. Antibiot. 30, 186.
244. Kinashi H., Otake N., Yonehara H., Sato S., Saito Y. (1975) Acta Crystallogr. B31, 2411.
245. Koenuma M., Kinashi H., Sato S., Saito Y. (1976) Acta Crystallogr. B32, 1267.
246. Omura S., Shibata M., Machida S., Sawada J. (1976) J. Antibiot. 29, 15.
247. Petcher T.J., Weber H.-P. (1974) J. Chem. Soc. D, Chem. Commun. 697.
248. Alleaume M., Busetta B., Farges C., Gachon P., Kergomard A., Staron T. (1975) J. Chem. Soc. D, Chem. Commun. 411.
249. Kinashi H., Otake N., Yonehara H. (1973) Tetrahedron Lett. 4955.
250. Riche C., Pascard-Billy C. (1975) J. Chem. Soc. D, Chem. Commun. 951.
251. Blount J.F., Westley J.W. (1975) J. Chem. Soc. D, Chem. Commun. 533.
252. Dorman D.E., Paschal J.W., Nakatsukasa W.M., Kuckstep L.L., Neuss N. (1976) Helv. Chim. Acta 59, 2625.
253. Steinrauf L.K., Pinkerton M., Chamberlin J.W. (1968) Biochem. Biophys. Res. Commun. 33, 29.

254. Kubota T., Matsutani S., Shiro M., Koyama H. (1968) Chem. Commun. 1541.
255. Pressman B.C., Haynes D.H. (1969) In: D.C. Tosteson, ed.: *The Molecular Basis of Membrane Function*. Prentice Hall, Englewood Cliffs, p. 211.
256. Henderson P.J.F., McGivan J.D., Chappell J.B. (1969) Biochem. J. *111*, 521.
257. Ashton R., Steinrauf L.K. (1970) J. Mol. Biol. *49*, 547.
258. Sumskaya L.V., Chekhlayeva N.M., Barsukov L.I., Terekhov O.P., Demin V.V., Shkrob A.M., Ivanov V.T., Ovchinnikov Yu.A. (1976) Bioorgan. Khim. (USSR) *2*, 351.
259. Pressman B.C., de Guzman N.T. (1974) Ann. N.Y. Acad. Sci. *227*, 380.
260. Berger J., Kachlin A.I., Scott W.E., Sternbach L.H., Goldberg M.W. (1951) J. Amer. Chem. Soc. *73*, 5295.
261. Pressman B.C. (1972) In: M.A. Mehlman, R.W. Hanson, eds.: *The Role of Membranes in Metabolic Regulation*. Academic Press, New York, N.Y., p. 149.
262. Johnson S.M., Herrin J., Lin S.J., Paul I.C. (1970) J. Amer. Chem. Soc. *92*, 4428.
263. Degani H., Friedman H.L., Navon G., Kosower E.M. (1973) J. Chem. Soc. D, Chem. Commun. 431.
264. Lindenbaum S., Sternson L., Rippel S. (1977) Chem. Commun. 268.
265. Westley J.W., Evans R.H., Blount J.F. (1977) J. Amer. Chem. Soc. *99*, 6057.
266. Shen C., Patel D.J. (1977) Proc. Natl. Acad. Sci. USA *74*, 4734.
267. Pressman B.C. (1973) Fed. Proc. *32*, 1698.
268. Westley J.W., Oliveto E.P., Berger J., Evans R.H., Glass R., Stempel A., Toome V., Williams T. (1973) J. Med. Chem. *16*, 397.
269. Schmidt P.G., Wang A.H.-J., Paul I.C. (1974) J. Amer. Chem. Soc. *96*, 6189.
270. Degani H., Friedman H.L. (1974) Biochemistry *13*, 5022.
271. Cornelius G., Gärtner W., Haynes D.H. (1974) Biochemistry *13*, 3052.
272. Haynes D.H., Pressman B.C. (1974) J. Membr. Biol. *16*, 195.
273. Degani H., Friedman H.L. (1975) Biochemistry *14*, 3755.
274. Patel D.J., Shen C. (1976) Proc. Natl. Acad. Sci. USA *73*, 1786.
275. Shen C., Patel D.J. (1976) Proc. Natl. Acad. Sci. USA *73*, 4277.
276. Reed P.W., Lardy H.A. (1972) J. Biol. Chem. *247*, 6970.
277. Reed P.W., Lardy H.A. (1972) In: M.A. Mehlman, R.W. Hanson, eds.: *The Role of Membranes in Metabolic Regulation*. Academic Press, New York, N.Y., p. 111.
278. Pfeiffer D.R., Reed P.W., Lardy H.A. (1974) Biochemistry *13*, 4007.
279. Chaney M.O., Demarco P.V., Jones N.D., Occolowitz J.L. (1974) J. Amer. Chem. Soc. *96*, 1932.
280. Chaney M.O., Jones N.D., Debono M. (1976) J. Antibiot. *29*, 424.
281. Anteunis M.J.O. (1977) Bioorg. Chem. *6*, 1.
282. Bottenstein J.E., deVellis J. (1976) Biochem. Biophys. Res. Commun. *73*, 486.
283. Wong D.T. (1976) FEBS Lett. *71*, 175.
284. Lindenbaum S., Sternson L., Rippel S. (1977) J. Chem. Soc. D, Chem. Commun. 268.

285. Duszynski J., Wojtczak L. (1977) Biochem. Biophys. Res. Commun. 74, 417.
286. Taylor D., Baker R., Hochstein P. (1977) Biochem. Biophys. Res. Commun. 76, 205.
287. Wulf J., Pohl W.G. (1977) Biochim. Biophys. Acta 465, 471.
288. Pfeiffer D.R., Lardy H.A. (1976) Biochemistry 15, 935.
289. Yang S.P., Gomperts B.D. (1977) Biochim. Biophys. Acta 469, 281.
290. Solomon A.K., Lionetti F., Curran P.F. (1956) Nature (London) 178, 582.
291. Rosano H.L., Duby P., Schulman J.H. (1961) J. Phys. Chem. 65, 1704.
292. Rosano H.L., Schiff H., Schulman J.H. (1962) J. Phys. Chem. 66, 1928.
293. Sears D.F., Schulman J.H. (1964) J. Phys. Chem. 68, 3529.
294. Schneider P.B., Wolff J. (1965) Biochim. Biophys. Acta 94, 114.
295. Feinstein M.B. (1964) J. Gen. Physiol. 48, 357.
296. Harris R.A., Farmer B. (1973) Lipids 8, 717.
297. Agate A.D., Vishniac W. (1972) Chem. Phys. Lipids 9, 247.
298. Tyson C.A., Zande H.V., Green D.E. (1976) J. Biol. Chem. 251, 1326.
299. Moore J.H., Schechter R.S. (1969) Nature (London) 222, 476.
300. Wojtczak L. (1974) FEBS Lett. 44, 25.
301. Blondin G.A. (1974) Ann. N.Y. Acad. Sci. 227, 392.
302. Blondin G.A. (1974) Biochem. Biophys. Res. Commun. 56, 97.
303. Blondin G.A. (1975) Ann. N.Y. Acad. Sci. 264, 98.
304. Carafoli E., Crovetti F. (1973) Arch. Biochem. Biophys. 154, 40.
305. Carsten M.E., Miller J.D. (1977) J. Biol. Chem. 252, 1576.
306. E.M. Southern, ed.: *The Prostaglandins. Clinical Applications in Human Reproduction*. Future Publ. Co. Inc., New York, N.Y., 1972.
307. Curtis-Prior P.B. (1976) *Prostaglandins. An Introduction to their Biochemistry, Physiology and Pharmacology*. North-Holland Publ. Co., Amsterdam.
308. Hendriks Th., Klompmakers A.A., Daemen F.J.M., Bonting S.L. (1976) Biochim. Biophys. Acta 433, 271.
309. Montal M., Darszon A., Trissl H.W. (1977) Nature (London) 76, 81.
310. Bolton J.E., Field M. (1977) J. Membr. Biol. 35, 159.
311. Smythies J.R., Benington F., Bradley R.J., Bridgers W.F., Morin R.D. (1974) J. Theor. Biol. 43, 29.
312. Finkelstein A., Rubin L.L., Tzeng M.-C. (1976) Science 193, 1009.
313. Villegas R., Villegas G.M., Barnela F.V., Racker E. (1977) Biochem. Biophys. Res. Commun. 79, 210.
314. Shamoo A.E., Goldstein D.A. (1972) Biochim. Biophys. Acta 472, 13.
315. Evtodienko Yu.V., Peshkova L.V., Shchipakin V.N. (1971) Ukrainian J. Biochem. (USSR) 43, 98.
316. Lehninger A.L. (1971) Biochem. Biophys. Res. Commun. 42, 312.
317. Gómez-Puyou A., De Gómez-Puyou M., Becker G., Lehninger A.L. (1972) Biochem. Biophys. Res. Commun. 47, 814.
318. Kimura T., Chu J.W., Mukai K., Ishizuka I., Yamakawa T. (1972) Biochem. Biophys. Res. Commun. 49, 1678.
319. Tashmukhamedov B.A., Gagelgans A.I., Mamatkubov Kh., Makhmudova E.M. (1972) FEBS Lett. 28, 239.

320. Makhmudova E.M., Gagelans A.I., Mirkhodzhaev U.Z., Tashmukhamedov B.A. (1975) Biofizika (USSR) *20*, 225.
321. Utsumi K., Oda T. (1974) In: M. Nakao, L. Packer, eds.: *Organization of Energy-transducing Membranes*. University Park Press, Baltimore, Md., p. 265.
322. Carafoli E., Gazzotti P., Vasington F.D., Sottocasa G.L., Sandri G., Panfili E., de Bernard B. (1972) In: G.F. Azzone, E. Carafoli, A.L. Lehninger, E. Quagliariello, N. Siliprandi, eds.: *Biochemistry and Biophysics of Mitochondrial Membranes*. Academic Press, New York, N.Y., p. 623.
323. Sottocasa G., Sandri G., Panfili E., de Bernard B., Gazzotti P., Vasington F.D., Carafoli E. (1972) Biochem. Biophys. Res. Commun. *47*, 808.
324. Carafoli E., Sottocasa G.L. (1974) In: L. Ernster, R. Estabrook, E.C. Slater, eds.: *Dynamics of Energy-transducing Membranes*. Elsevier, Amsterdam, p. 455.
325. Prestipino G., Ceccarelli D., Conti F., Carafoli E. (1974) FEBS Lett. *45*, 99.
326. Carafoli E. (1976) In: L. Parker, A. Gómez-Puyou, eds.: *Mitochondria Bioenergetics, Biogenesis and Membrane Structure*. Academic Press, New York, N.Y., p. 47.
327. Carafoli E., Crompton M. (1976) In: A. Martonosi, ed.: *The Enzymes of Biological Membranes*, Vol. 3. Plenum Press, New York, N.Y., p. 193.
328. Pardee A.B. (1968) Science *162*, 632.
329. Heppel L.A. (1971) In: L.I. Rothfield, ed.: *Membrane Structure and Function*. Academic Press, New York, N.Y.
330. Oxender D.L. (1972) Ann. Rev. Biochem. *41*, 777.
331. Boos W. (1974) Ann. Rev. Biochem. *43*, 123.
332. Antonov V.K., Alexandrov S.L. (1977) Bioorgan. Khim. (USSR) *3*, 581.
333. Oxender D.L., Quay, S. (1975) Ann. N.Y. Acad. Sci. *264*, 358.
334. Pardee A.B. (1966) J. Biol. Chem. *241*, 5886.
335. Imagawa T. (1972) J. Biochem. *72*, 911.
336. Imagawa T., Suzuki S., Tsugita A. (1972) J. Biochem. *72*, 927.
337. Medveczky N., Rosenberg H. (1969) Biochim. Biophys. Acta *192*, 369.
338. Medveczky N., Rosenberg H. (1970) Biochim. Biophys. Acta *211*, 158.
339. Piperno J.R., Oxender D.L. (1966) J. Biol. Chem. *241*, 5732.
340. Winkler H.H., Wilson T.H. (1967) Biochim. Biophys. Acta *135*, 1030.
341. Scarborough G.A., Rumley M.K., Kennedy E.P. (1968) Proc. Natl. Acad. Sci. USA *60*, 951.
342. Yariv J., Kalb A.J., Katchalski E., Goldman R., Thomas E.W. (1969) FEBS Lett. *5*, 173.
343. Langridge R., Shinagawa H., Pardee A.B. (1970) Science *169*, 59.
344. Boos W., Gordon A.S. (1971) J. Biol. Chem. *246*, 621.
345. Ovchinnikov Yu.A., Aldanova N.A., Grinkevich V.A., Arzamazova N.M., Moroz I.N. (1977) FEBS Lett. *78*, 313.
346. Ovchinnikov Yu.A. (1978) In: *Frontiers in Physicochemical Biology*. Academic Press, Inc., New York, N.Y., in press.
347. Hogg R.W., Hermodson M.A. (1977) J. Biol. Chem. *252*, 5135.
348. Quiocho F.A., Gilliland G.L., Phillips G.N., Jr. (1977) J. Biol. Chem. *252*, 5142.

Yu.A. Ovchinnikov and M.N. Kolosov (eds.) Frontiers in Bioorganic Chemistry and Molecular Biology (©) 1979, Elsevier/North-Holland Biomedical Press

CHAPTER 9

The β-replacing and α,β- eliminating pyridoxal-*P*-dependent lyases: enantiomeric cycloserine pseudo-substrates and the catalytic mechanism

A.E. BRAUNSTEIN, E.V. GORYACHENKOVA, R.A. KAZARYAN, L.A. POLYAKOVA and E.A. TOLOSA

Introduction

The general theory of amino acid transformations by pyridoxal phosphate (PLP) and by PLP-containing enzymes was formulated 25 years ago by A.E. Braunstein and M.M. Shemyakin [*1,2*]. Soon after, E.E. Snell and his associates in the USA independently proposed a very similar theory. At that time we further found, that pyridoxal phosphate takes part in the biosynthesis and metabolic conversions of cysteine (desulfuration, transsulfuration etc. [2,*4,5*]). These and other reactions of biologically important α-amino acids with an electrophilic β or γ substituent (X) are catalyzed by a subgroup of widespread PLP enzymes classified as lyases (EC, class 4), effecting, more or less selectively, elimination and/or replacement of the X-substituent according to equations I–IV.

α,β-Elimination:

$$X\overset{\beta}{C}HR{\cdot}\overset{\alpha}{C}H\overset{+}{N}H_3{\cdot}COO^- + H_2O \rightleftharpoons XH + NH_4^+ + R\overset{\beta}{C}H_2{\cdot}\overset{\alpha}{C}O{\cdot}COO^- \qquad [I]$$

β-Replacement:

$$X\overset{\beta}{C}HR{\cdot}\overset{\alpha}{C}H\overset{+}{N}H_3{\cdot}COO^- + YH \rightleftharpoons XH + Y\overset{\beta}{C}HR{\cdot}\overset{\alpha}{C}HNH_3{\cdot}COO^- \qquad [II]$$

β,γ-Elimination:

$$\overset{\gamma}{X}CHR\cdot\overset{\beta}{C}H_2\cdot\overset{\alpha}{C}H\overset{+}{N}H_3\cdot COO^- + H_2O \rightarrow XH + NH_4^+ + \overset{\gamma}{R}CH_2\cdot\overset{\beta}{C}H_2\cdot\overset{\alpha}{C}O\cdot COO^-$$

[III]

γ-Replacement:

$$\overset{\gamma}{X}CHR\cdot\overset{\beta}{C}H_2\cdot\overset{\alpha}{C}H\overset{+}{N}H_3\cdot COO^- + YH \rightleftharpoons XH + \overset{\gamma}{Y}CHR\cdot\overset{\beta}{C}H_2\cdot\overset{\alpha}{C}HNH_3\cdot COO^-$$ [IV]

In the theories of pyridoxal catalysis developed by the Soviet and American [3,6] scientists, the mechanisms of these reactions were interpreted in similar but not quite identical ways.

Both the Soviet school [1,2,4,5] and the American authors [3,6–9] assume that in all type I–IV reactions, the first step in the action of PLP-dependent lyases is dissociation of the α-hydrogen in the enzyme-PLP-substrate aldimines ('external' Schiff bases [2,5]). The lyases often display group specificity towards the amino acid substrates and the cosubstrates, and some of them are plurifunctional, catalyzing more than one of the I–IV type reactions, a fact which had an influence on concepts concerning the reaction mechanism.

Snell and his school [3,6–9] assume that with all PLP lyases dissociation of the α-H atom in the substrate-PLP-aldimine is followed by tautomeric conversion to the PMP-ketimine. We share this view with regard to the γ-elimination and replacement reactions III and IV. Our view are also largely in accord concerning the mechanism of type (I) or type (I + II) reactions, catalyzed by α,β-eliminating or ambifunctional lyases [14]. These mechanisms are illustrated in Fig. 1 (cf. the similar, more detailed scheme of Snell [20] for bacterial tryptophanase).

The mechanism shown in Fig. 1 involves formation of α,β-unsaturated coenzyme-substrate imines (pyridoxylidene aminoacrylates,3') via the PMP-ketimines (2) and their transformation involving elimination of α-H and of the β-X substituent. In the α,β-elimination reactions (equation I), the intermediate unsaturated imines are non-enzymically hydrolyzed to NH_4^+, α-keto acid, and free PLP enzyme. In the β-replacement (type II) reactions, nucleophilic addition of a molecule of replacing agent YH (the cosubstrate) to the double bond of the $\Delta^{\alpha,\beta}$-unsaturated (aminoacrylate-coenzyme) imine (Michael reaction) occurs first, and this is followed by hydrolysis of the saturated Schiff base thus formed to the new amino acid, $YCHR \cdot CH\overset{+}{N}H_3 \cdot COO^-$ (4) and the free PLP enzyme. Such a reaction sequence has been substantiated for the β-eliminating (type I) lyases and the ambifunctional (types I + II) β-lyases by an ample body of indirect evidence [7,10] including: (a) rapid isotopic exchange of the α-H-atom

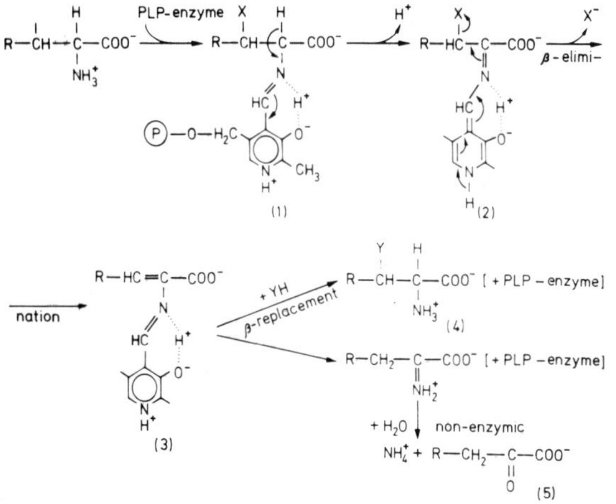

Fig. 1. Mechanism of α,β-elimination and β-replacement catalyzed by pyridoxal phosphate dependent lyases (according to Davis and Metzler [7]).

and the electrophilic β-substituent X (probably fixed in *trans* orientation, see below); (b) appearance in the enzyme-substrate spectra of long-wave absorption bands ($\lambda_{max} \geqslant 490-500$ nm), characteristic of quinonoid tautomers or of ($\Delta^{\alpha,\beta}$ or $\Delta^{\alpha\text{-C-N}}$) unsaturated imine intermediates with an extended conjugated π-system; (c) typical nucleophilic Michael addition to *N*-ethylmaleimide of the double-bonded β-carbon of the coenzyme-iminoacrylate complex in the enzyme's active site, to form a chiral adduct of the inhibitor, with suppression of α-keto acid (5) formation as first described by Flavin et al. [8]. The ambivalent lyase-catalyzed β-replacement reactions (I + II) in the steady state display Cleland [21] 'Ping Pong Bi Bi' kinetics:

$$\underset{\downarrow}{A^1} \quad \underset{\uparrow}{XH} \quad \underset{\downarrow}{YH} \quad \underset{\uparrow}{A^2}$$

$$E \rightarrow E{=}A \rightarrow E{=}A^1(\Delta^{\alpha,\beta}) \rightarrow E{=}A^2 \rightarrow E$$

(E – enzyme, A – amino acid).

In the case of α,β-elimination, reaction (I) proceeds by the 'Uni Ordered Tri' mechanism [21].

During the last few years, a subgroup of PLP-dependent lyases, selectively catalysing type II β-replacement reactions (see Table 1, enzymes 1–4) have been investigated in detail in our laboratory. When highly purified, these enzymes do not promote type I elimination under whatever conditions. Snell and his followers, not referring to this lyase subgroup in particular, assumed for all PLP-dependent β-replacement reactions a mechanism of the type shown in Fig. 1.

On the basis of abundant evidence and the fact that none of the criteria listed above is applicable to the four purified exclusively β-replacing PLP-dependent lyases prepared in our laboratory (see [4,5,10–14] and also [2,6–10,14–20]), we arrived at the conclusion that these enzymes must have a catalytic mechanism essentially different from that of all other types of β- and γ-specific lyases.

As early as in 1952–53, Braunstein and Shemyakin, taking into account the supplementary inductive effects of the electrophilic β-substituents, X, and of the cosubstrates, YH, proposed that α-H exchange and β-X elimination or replacement in the amino acid substrate molecules can proceed, particularly in the lyase active site, directly via the β-substituted PLP-aldimines, without intermediate tautomeric PMP-ketimine formation (see reference [2], schemes I and K on pp. 164–168). We had also then suggested a scheme alternative to K (see [2], scheme L) for the β-substitution reactions which moreover bypassed formation of intermediate unsaturated (coenzyme-aminoacrylate) imines. The validity of this mechanism, corresponding in essential features to the scheme reproduced on Fig. 2, is supported by theoretical considerations and by the evidence we and our associates (Nguyen Dinh Lac, I. Willhardt, R.N. Maslova, L.L. Yefremova, T.N. Akopyan, A.G. Rabinkov and others) have accumulated in studies of the four exclusively β-replacing lyases (Table 1, 1–4), and also of a few β-specific lyases (5–8) of other types, investigated for comparison.

Recently, in collaboration with L.V. Kozlov, we have analyzed the steady

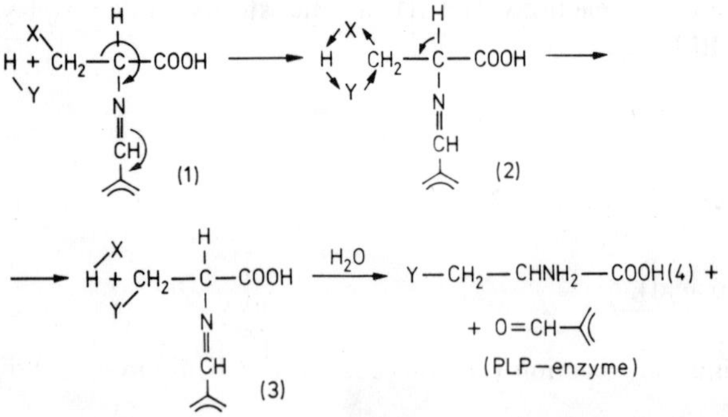

Fig. 2. Mechanism for β-replacement reactions catalyzed by pyridoxal phosphate dependent lyases (Scheme L of Braunstein and Schemyakin [2]).

Table 1. The PLP-dependent lyases studied [14].

Enzymes [a] (classification [10], name, biological source)	Types of reactions catalyzed	Primary substrates (β-substituted L-α-amino acids)	Replacing agents (cosubstrates)
A,2,b. β-Replacing			
1. Cysteine lyase (chicken embryo yolk sac)	II	Cys	HSO_3^-, AlkSH, Cys, H_2S
2. Serine sulfhydrase (chicken liver, yeast)	II	Ser, Cys, Ser(OAcyl), Cys(SAlk), Ala(Cl), Ala(CN)	Hcy, AlkSH, H_2S, $NH_2(CH_2)_2SH$, $HO(CH_2)_2SH$
3. Cystathionine-β-synthase (mammalian liver, microorganisms; allelosyme [27] of lyase 2)	II	Ser, Cys, Ser(OAcyl), Cys(SAlk), Ala(Cl), Ala(CN)	Hcy, AlkSH, H_2S, $NH_2(CH_2)_2SH$, $HO(CH_2)_2SH$
4. β-Cyanoalanine synthase (lupine seedlings)	II	Cys, Ala(Cl), Ala(SCN)	HCN, H_2S, MeSH
A,2,a. α,β-Eliminating	I		
5. Alliinase (garlic)		Alliin (and its analogs)	
6. Serine dehydratase (rat liver)	I	Ser, Thr, *erythro*-β-HO-Asp	
A,2,c. α,β-Eliminating and β-Replacing (ambifunctional)			
7. Tryptophanase (E. coli and other microorganisms)	I, II	Trp, Ser, Cys, Cys(SAlk)	3- and 5-Alk-Ind, H_2O, H_2S, AlkSH,
A,1,d. β,γ- and α,β-specific (polyfunctional)			
8. γ-Cystathionase [b] (liver)	I, II(?), III, IV	Cystathionine, Hse, Cys, C̄ys C̄ys	[b]

[a] Enzymes 1—5 and 8 were purified to 95—100% homogeneity by techniques developed in our laboratory.

[b] Evidence concerning reactions of types II and IV is scanty.

state kinetics of the β-replacement reaction using serine sulfhydrase (2) and β-cyanoalanine synthase (4) as representative lyases of this subgroup. The results indicated that the catalytic mechanism of this subgroup (as opposed to that of ambivalent lyases capable of catalyzing α,β-elimination as well) is essentially in agreement with the scheme shown in Fig. 2. The reactions proceed *via* intermediate formation of a ternary amino-substrate-PLP-enzyme-cosubstrate complex (E=A^1 · YH), which further undergoes 'Random Bi Ordered Bi' transformation (according to the Cleland system).

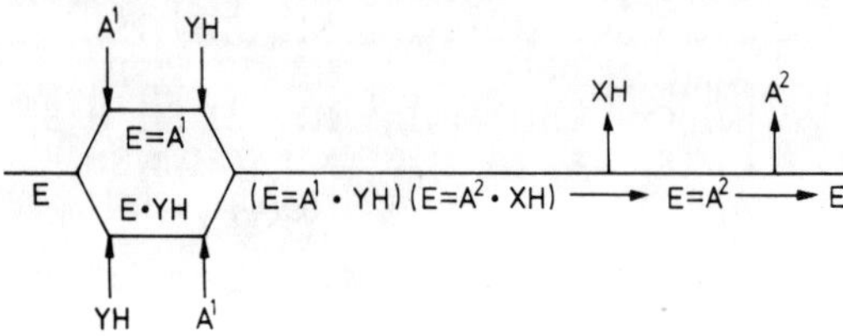

```
         A¹   YH                    XH    A²
          \  /                       |    |
         E=A¹                        ↓    ↓
E                (E=A¹ · YH)(E=A² · XH) ⟶ E=A² ⟶ E
         E·YH
          /  \
         YH   A¹
```

A, amino acid   E, enzyme

Interaction of the PLP-dependent lyases with enantiomeric cycloserines

Under this heading we discuss results obtained in applying one of the approaches utilized in our laboratory for the mechanistic study of PLP enzymes, namely the use of the enantiomers of 4-aminoisoxazolid-3-one to elucidate the differences in the modes of action of the eliminating and β-replacing PLP-dependent lyases (Table 2).

The naturally occurring enantiomer, the antibiotic cycloserine is the cyclic alkoxamide of 3-aminooxy-D-alanine. It is, accordingly, designated as D-cycloserine, and its enantiomer as L-cycloserine. These compounds are sterically rigid structural analogs (and biological antagonists) of the corresponding alanine enantiomers, and the 5-substituted D- and L-cycloserine derivatives (5-R-cycloserines) are cyclic analogs of the higher α-amino acids.

A. α-Amino acids

B. Cycloserine and its 5-substituted derivatives

For example, substitution of 5-H in cycloserine by a carboxyethyl group (R = CH_2CH_2COOH) results in a mixture of *threo-* and *erythro*-DL-α-cycloglutamates — diastereomeric forms of the cyclic DL-glutamic acid analog (R.M. Khomutov et al. [15–17]).

Compounds of this series are biologically active as inhibitory pseudosubstrates, suppressing the enzyme-catalyzed reactions of the structurally and configurationally cognate natural amino acids. Thus, the antibacterial properties of the natural antibiotic D-cycloserine were found by a number of authors to be the result of potent inhibition of bacterial enzymes catalyzing the metabolic transformations of D-alanine, among them the PLP-dependent D-transaminases and racemases [2,19], and of the suppression of enzyme-catalyzed incorporation of D-alanine into the bacterial cell-wall peptidoglycans (Strominger et al.).

In animal and higher plant tissues the more potent agents are the L-enantiomers of cycloserine and its derivatives, which act as inhibitors of various PLP enzymes transforming L-amino acids.

In a number of reviews [10,14,19] Braunstein has discussed the well-known theory of M.Ya. Karpeisky, R.M. Khomutov, E.S. Severin et al. [15–18] regarding the mechanism of irreversible inactivation of transaminases by these and related compounds. According to these authors, L-aminoisoxazolidones act as pseudosubstrates binding the PLP in the active center of the transaminases to form 'external' aldimines (Fig. 3,2). These are then converted by interaction with catalytic groups of the enzyme into unstable PMP-ketimines of the isoxazolid-3-ones (Fig. 3,3'). Such ketimines and their degradation products and (after acid hydrolysis) pyridoxamine can be isolated from the deproteinized reaction mixture.

Formation of the ketimine is essential for the irreversible enzyme inactivation because imines of the 3' type readily undergo ring opening and covalently block (acylate) a nucleophilic group B in the active center to form the inactive complex EI_{335}. The inhibition is atypical or is absent altogether if the experimental conditions are unfavorable for the formation of covalently bound PMP-ketimine complexes, as for example, when D-aminoisoxazolidones or L-enantiomer derivatives of unsuitable configuration are used [15–17], or else when the PLP enzymes are of the resistant type, such as glutamate α-decarboxylase or certain β-specific lyases [10,14,18]. When inhibition was sometimes observed with such systems (at times with very high K_i levels), this was due to decyclization of the aminoisoxazolidone (because of prolonged storage or of a lengthy preincubation period) to O-substituted hydroxylamines which are potent inhibitors by virtue of oxime formation with the PLP in the active center (cf. the first footnote to Table 2).

It has been suggested [10] that typical irreversible inhibition by aminoiso-

Table 2. Inhibitory effects of the enantiomers of cycloserine upon PLP-dependent enzymes, including lyases [a].

Enzyme (biological source)	Degree of purification	I_{50} (mM) cycloserine			Preincubation conditions			References
		L	DL	D	Time min	pH	PLP concentration M	
a. Ala : Glu transaminase (pig-heart cytosol)	partial	0.02	0.05	0.8	5	7.4		16
b. Asp : Glu transaminase (pig-heart cytosol)	partial	0.7	3.5	>>10	5	8.3		15
same	same		0.16		30	5.8		16
c. Glu α-decarboxylase	partial and ⩾90%	no inhibition [b]			10	4.6		10, 14, 18 [c]
1. Cysteine lyase (chicken embryo yolk sec)	~90%	no inhibition [b]			15	8.2	0	10, 14
2. Serine sulfhydrase (chicken liver, yeast)	>95%	no inhibition [b]			20	8.2	0	14
		inhibition [b]			20	8.2	10^{-5}	10, 14

Enzyme	Inhibition							
5. Alliinase (garlic)	>90%	0.04	~0.1		0.3	20	7.5	10^{-5}
6. Serine dehydratase (rat liver)	same	0.008	≤0.05		0.08	20	7.5	0
	partial		<<0.5		0.5	20	8.0	0
7. Tryptophanase (E. coli)	partial		0.07			10	8.3	10^{-5}	10
8. γ-Cystathionase (rat liver)	>95%	0.01	~0.07	~0.8		20	8.0	10^{-5}

[a] The D-isomer of cycloserine specifically and irreversibly inhibits PLP enzymes capable of acting upon D-amino acids, such as D-transaminases, racemases, serine hydroxymethylase. PLP enzymes resistant to pure D- or L-cycloserine may undergo considerable inactivation (non-specific and mostly reversible), if cycloserine preparations are used following prolonged storage or long-lasting pre-incubation with the enzyme solution. This is caused by decyclization of the isoxazolidone ring with formation of aminooxy compounds, intensely inhibitory owing to the formation of substituted oximes of pyridoxal phosphate. Most of the PLP enzymes are inhibited by NH_2OH and its O-substituted derivatives, especially those structurally analogous to substrates with K_i values of $10^{-5}-10^{-6}$ M. The Table is reproduced from Ref. [14] and supplemented by newer data of the present authors and others, relating to enzymes **2–6** and **8** (1977).

[b] Cycloserine concentrations up to 20–50 mM were preincubated with the β-replacing lyases in buffered solution in the presence of an adequate cosubstrate, e.g. β-mercaptoethanol [12].

[c] According to Turano (personal communication), cycloserine does not inhibit DOPA decarboxylase either.

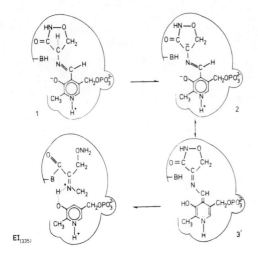

Fig. 3. Mechanism of irreversible inactivation of transaminases by cycloserine [16,17,19].

oxazolidones with substrate-like configuration (formation of EI_{335} complexes) may be characteristic only of such PLP enzymes, in the normal catalytic mechanism of which an obligatory step is PMP-ketimine formation. The data presented in Table 2 and discussed below appear to support this view.

Unfortunately, in most studies, especially of the early period, enzymologists compared the effect of D-cycloserine with that of racemic cycloserine and of its stereoisomeric derivatives rather than of the L-enantiomers. (Also, old preparations contaminated with decyclized products were often employed.) Because of considerable differences in kinetic constants and in the inhibitory mechanisms of the isomeric forms and their degradation products, these faulty experimental conditions present serious stumbling blocks to correct interpretation of the results reported.

This motivated us to repeat and further extend our earlier experiments, using highly purified representatives of the various subgroups of eliminating and replacing lyases, and fresh solutions of pure L- and D-cycloserine.

The latter were obtained from an aqueous solution of racemic cycloserine by fractional precipitation of the diastereoisomeric D-tartrates; D-cycloserine D-tartrate crystallized first, and the D-tartrate of L-cycloserine was recovered from the mother liquor. The free cycloserine enantiomers were obtained by column chromatography of the crude enantiomer D-tartrates on ion-exchange resins and repeated crystallization.

Table 3. Mechanism of inactivation of eliminating lyases by L- (or DL-) and D-cycloserine (cSer).

Sample No.	Fraction under study [a]	Activity recovery [b]							
		Allinase [c]				Serine dehydratase [d]		γ-Cystathionase [d]	
		PLP absent		PLP present		DL-cSer	D-cSer	L-cSer	D-cSer
		L-cSer	D-cSer	L-cSer	D-cSer				
1 [e]	Eluted protein	0	0	19	18	0	0	0	0
2 [f]	Eluted protein plus 25 mg PLP	49	52	90	100	41	90	53	90

[a] Each lyase was preincubated 30 min with cycloserine; no residual activity was observed after preincubation followed by collection of the required major protein fraction by means of gel-filtration through Sephadex G-50 or G-75.
[b] The preincubation samples contained 10^{-6} M enzyme, 10^{-2} M cycloserine and either 10^{-5} M PLP or no added PLP. The examined enzymes originally had the following specific activities (μmol/hr per 1 mg of the preparation): allinase — 3600; serine dehydratase — 35, γ-cystathionase — 200–300. Activity recovery is expressed in percentage of the original specific activity of the given preparation.
[c] The enzyme was preincubated with cycloserine at pH 7.5 and 23°.
[d] The enzyme was preincubated with cycloserine in 0.1 M potassium phosphate buffer at pH 8.1, and 30°.
[e] These data demonstrate absence of homodromic reversibility, i.e. of dissociation of E-I complex upon gel-filtration with release of active holoenzyme.
[f] The data show the extent of reassociation of apo-enzyme with PLP to active holoenzyme under the stated conditions.

Table 2 compares the inhibitory effects of the enantiomeric cycloserines upon the activity of the various PLP-dependent lyases and certain other PLP enzymes. Four lyases, exclusively catalyzing only the β-replacement reactions (1—4), and amino acid α-decarboxylases proved completely resistant to both D- and L-cycloserine. On the contrary, PLP-dependent lyases of other types (5—8 in Table 1) are under certain conditions (see Tables 2 and 3) reversibly inhibited by L-cycloserine with formation of PMP-imines [10,14].

In experiments with serine sulfhydrase (2) or cyanoalanine synthase (4), no significant shift of the chromophore absorption peak (of relatively low intensity in some β-replacing lyases) is observed even after lengthy incubation with high concentrations (10^{-2} M) of DL-cycloserine at pH 8.2. After mild acid denaturation and heating of the non-protein fraction of the mixture to boiling in 1 N HCl, paper chromatography revealed the presence of pyridoxal and PLP, with no trace of pyridoxamine or PMP (Fig. 4). These lyases were treated with cycloserine and then reduced with sodium borotritide; when such mixtures were gel-filtered through Sephadex G-25, the radioactivity was largely retained in the protein fraction. The radioactivity remained predominantly there also after acidification of the protein solution to pH 5.0 and recovery of the denatured protein (presumably, it was the 'internal' PLP-lysine aldimine that was reduced). However, ^{3}H-labeled pyridoxylcycloserine and its decomposition products were

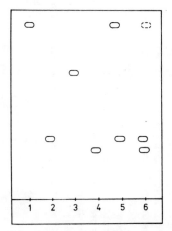

Fig. 4. Analysis of the cycloserine-coenzyme complexes in serine sulfhydrase and γ-cystathionase. Paper chromatography on FN-16 (Filtrak) in the solvent system: methyl ethyl ketone, propionic acid, water (15 : 5 : 6).
(1) pyridoxal; (2) pyridoxal phosphate; (3) pyridoxamine; (4) pyridoxamine phosphate; (5) serinesulfhydrase complex; (6) γ-cystathionase complex.

detected in the eluate (Fig. 5); this indicates that PLP-cycloserine aldimine formation is possible in β-replacing lyases, but only to a restricted extent, because of the low affinity of cycloserine for the active site in these enzymes.

Completely different results were obtained with practically pure, homogeneous preparations of eliminating lyases, namely, alliinase, γ-cystathionase (using homoserine as the substrate), and partially purified serine dehydratase. These enzymes were highly sensitive to inhibition by L-cycloserine (Table 2).

Alliinase (5) in a PLP-free solution (i.e. with an incompletely coenzyme-saturated active site) was inhibited by L-cycloserine with an I_{50} of approx. $8 \cdot 10^{-6}$ M while the affinity of the enzyme for the D-enantiomer was about 10-fold lower. The protein fraction obtained on gel-filtration of the inactive L-cycloserine-enzyme complex through Sephadex G-25 contains the apoenzyme in native form. Its catalytic activity is considerably regenerated if PLP is immediately added to the eluate (Table 3). In the absence of PLP, however, apo-alliinase is very unstable and rapidly loses its capacity for reactivation. The non-protein fraction contains coenzyme-inhibitor imines which decompose on mild acid hydrolysis, yielding mainly PMP and free pyridoxamine. Similar results

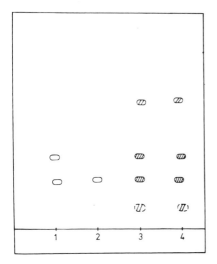

Fig. 5. Chromatographic analysis of cycloserine-coenzyme complexes in β-replacing lyases (after sodium borotritide reduction). Chromatographic conditions the same as in Fig. 4.
(1) N-pyridoxylcycloserine; (2) N^{α}-pyridoxyl-β-amino-oxylalanine; (3) β-cyanoalanine synthase complex; (4) serinesulfhydrase complex.
Hatched spots indicate ^{3}H-labeled substances.

were obtained in experiments with moderately purified liver serine dehydratase (**6**; an α,β-eliminating enzyme) and with γ-cystathionase (**8**) (Fig. 4); the latter is a plurifunctional PLP lyase. Like alliinase it possesses high L-cycloserine affinity ($I_{50} = 10^{-5}$ M); affinity for the D-isomer was about 70 times less (Table 2). In the case of γ-cystathionase, the chromophore of which has absorption and positive CD maxima at 427 nm, treatment with L- or D-cycloserine (with 10^{-5} M PLP added to the buffer solution) produced a catalytically inactive enzyme-cycloserine complex with a dichroic λ_{max} at 340 nm. This EI_{340}-complex, unlike similar complexes (EI_{335}) of transaminase inactivated with D-4-aminoisoxazolid-3-one, apparently has no stable covalent bond with the inhibitor; gel-filtration yields native apo-cystathionase, readily reactivated with pyridoxal phosphate, and also low-molecular products in which the coenzyme is in the form of PMP-ketimines (Table 3 and Fig. 4).

On the basis of earlier and the above described data the following conclusions can be made:

(a) In β-replacing lyases (for instance, **2** and **4**), the equilibrium between the 'internal' (PLP-lysine) and 'external' (PLP-cycloserine) aldimines in the active site is shifted towards the former (or to PLP-substrate aldimines, in the presence of appropriate amino acid substrates). For this reason the demonstrable affinity of the holoenzymes for L-cycloserine is very low, and much lower still for the D-isomer. Such lyases gave no evidence of aldimine to ketimine rearrangement of the cycloserine adducts. This is substantiated by the absence of ^{3}H-incorporation into the holoenzyme and into the excess cycloserine on incubation of lyase **2** in a ^3H_2O-containing medium in the presence of an appropriate cosubstrate (for instance, mercaptoethanol); the latter is required by β-replacing lyases for labilization and exchange of α-H in the amino acid substrates and in quasisubstrate inhibitors (see [*12*], and Table 4).

(b) Interaction of L-cycloserine with α,β-eliminating lyases, for example, alliinase (**5**), serine dehydratase (**6**), and γ-cystathionase (**8**), results in the formation of PMP-ketimines containing degradation products of the inhibitor, as in the case of transaminases (and probably of other PLP enzymes susceptible to irreversible inactivation by cycloserine).

The formation of PMP-inhibitor imines is manifested, as in the case of γ-cystathionase, for example, by a shift in the dichroic absorption band of the enzyme from λ_{max} 427 nm to λ_{max} 340 nm. It should be borne in mind, however, that this enzyme and other β-eliminating PLP lyases form iminoxy compounds with hydroxylamine or its O-substituted derivatives, the absorption and CD bands of which have λ_{max} near 340 nm, rather than in the 380–390 nm region as in the case of transaminases. The λ_{max} 340 oximes are not easy to distinguish from enzyme-bound coenzyme-inhibitor imines, which also display CD and absorption maxima at 340 nm. Hence, it is possible that inactive complexes of eliminating lyases modified with cycloserine might contain oximes of the enzyme with decyclization products of the inhibitor, and thus could be regenerated to active enzyme by added pyridoxal phosphate.

Our experiments have shown, however, that the PMP-ketimines produced by L-cycloserine in the active site of eliminating lyases (in contrast to transaminases [15–17]) do not manifest stable covalent modification (acylation by decyclized inhibitor) of an essential nucleophilic group of the enzyme in proximity to the active center. Native apoenzyme can, therefore, be readily liberated (by gel-filtration, for instance) from the EI_{340} cycloserine complexes of lyases 5, 6 or 8 and reactivated by adding pyridoxal phosphate.

The β-replacing lyases (1–4) proved insensitive to the substrate analog, L-cycloserine, and incapable of converting its PLP-aldimine to PMP-ketimines. Possibly this is due to the same cause as their inability to catalyze β-elimination reactions and the isotopic exchange of α-H in most of the structurally analogous amino acid pseudosubstrates, even in the presence of the cosubstrate YH. For the α-H exchange, the presence of appropriate cosubstrate is required even in the case of the most suitable amino acid substrates [12–14].

Distinctive molecular and catalytic properties of various subgroups of PLP-dependent lyases

One such cause (see above) is a difference in conformation of the substrate molecule bound in the active site of the different lyases, as we have shown albeit by indirect experimental evidence. We assume [10–14] that in elimination-specific β-lyases the amino acid substrate is fixed in a conformation with the β-substituent X-oriented *trans* to the α-H (*cis* to the NH_2 group) as in most chemical and enzymic 1,2-elimination reactions. Such geometry is in accord with the sensitivity of eliminating and plurifunctional lyases to inhibition by 1,2- and 1,3-aminothiols (in particular, to substrate-analogous 3- or 4-mercapto-2-amino acids such as penicillamine, homocysteine, cysteine, etc.) which form stable, inactive heterocyclic adducts (substituted thiazolidines and thiazanes) with the HCO group of PLP (cf. [24,25]).

If in the β-replacing lyases the β-substituent groups of the substrate PLP aldimines are fixed in the opposite orientation (namely, *cis* to the α-H atom and *trans* to the NH_2 group), this could account both for the failure of such lyases to promote α,β-elimination and for their resistance to inhibition by substrate-analogous aminothiols [10–14] (cf. Table 3) which we have demonstrated in conformity with the above postulate.

Possibly the β-specific lyases require no catalysis by a proton acceptor group of the protein to dissociate the α-H atom from the substrate PLP-aldimine by virtue of additional decrease in electron density of the α-C atom in the amino acid substrates due to the internal inductive effect of the *trans* oriented electro-

Table 4. Distinctive features of β-replacement-specific and other PLP-dependent lyases [a].

A	B
Criteria	Traits of exclusively β-replacing lyases
Reaction types (I-IV) and reversibility	$\overrightarrow{\text{II}}$ (lyases **1** and **4**), $\rightleftharpoons$II (lyases **2** and **3**)
Stationary-kinetic mechanism according to the Cleland system [*21*]	'Random Bi Ordered Bi' (**2, 4**)
Amino acid substrates	Cys, Ala(Cl) (**1** and **4**); Ala(SCN) (**4**); Ser, Cys, Cys(SAlk) (**2** and **3**)
Replacing agents (YH) in reactions of types II (resp. I + II) and IV	HSO_3^-, AlkSH, H_2S, Cys(**1**); Hcy, AlkSH, H_2S, H_2O (**2, 3**); CNH, MeSH, HS · Et · NH_2 (**4**)
Isotopic exchange of H-atoms and β-substituent	without YH: β-H exchange, practically no α-H exchange; with YH: no β-H exchange; rate of α-H exchange in non-reacted (excess) substrate is approximately equivalent to the initial rate of reaction II; β-substituent exchange is insignificant (lyases **1, 2** and **4**)
Effects of 3- and 4-mercapto-2-amino acids (penicillamine, cysteine, homocysteine, etc.)	no inhibition (**1–4**)

negative β-substituent X. One might then suppose that such an inductive effect is apparently not sufficient when X is *cis* to the α-H atom. If so, the α-deprotonation might require additional external induction by a suitably located molecule (or anion) of the replacing agent YH. This could increase polarization of the

Traits of eliminating and plurifunctional ...ases	D References and comments
(5 and 6), $\overrightarrow{\text{I}}, \overrightarrow{\text{II}}$ (7), $\overrightarrow{\text{I}}, \overrightarrow{\text{III}}, \overleftrightarrow{\text{II}}? \overrightarrow{\text{IV}}$ (8)	B [10–14]; C [6–9,20]; reactions I could with high concentrations of endproducts be reversed in the cases of tryptophanase (7) and β-tyrosinase [20];
reactions I: 'Uni sequential Tri' reactions II: 'Ping Pong Bi Bi'	B: our findings (1977); C: according to the scheme on Fig. 1
...liin and its analogs (5); Ser, Thr(6); ...p, Ser, Cys, Cys(SAlk), 8-Me- and ...Me-Trp, etc. (7); Hcy, Hse, Cys, ...s-Cys-, Hse(OAcyl)? (8)	B [10,11,13,14]; C [6–8,20]
...S, RSH, H$_2$O, Ind, (Alk)Ind (7, in actions II)	B [11,13,14]; C [7,8,20]
...ase 7: α-H exchange is rapid in sub...ates and moderately fast in inhibitory ...alogs; ...I exchange is moderate (chiral in ...ino acids with 4 or more C-atoms) ...ase 8: α-H exchange is rapid, β-H ...change is extremely rapid (chiral in ...substrates and quasi substrates)	B [12,14]: YH must be adequate for the given lyase; C [7–9,20]
...ibition largely competitive, owing ...formation of heterocyclic adducts ...h PLP in the active site; the EI com...unds readily undergo resolution with ...ase of apoenzyme (5–8 and others), [24]	Interpretation [10,11,14]: in β-replacement-specific lyases the substrate imines are fixed in the active site with β-X in *cis* orientation versus α-H; in eliminating lyases the orientation is *trans;* ambivalent lyases apparently have freedom of rotation around the α-C-β-C bond. This accounts for the failure of lyases 1–4 to catalyze α,β-elimination, as well as for their refractoriness to inhibition by aminothiols (resp. mercaptoamino acids)

Table 4 (continued)

A	B
Criteria	Traits of exclusively β-replacing lyases
Spectra of absorption and circular dichroism (CD) with $\lambda_{max} \geqslant 480-500$ nm	Spectral bands in this range are not observ(ed) in S · E and quasi S · E-complexes of lyase 1—4
Interactions with L- (or DL-) and D-cycloserine	Fail to inhibit lyases 1—4. Weak reversible binding of cycloserine in the form of PLP aldimines can be demonstrated at high cy(clo)serine concentration (no less than 0.1 M). Acidic denaturation of the cSer-enzyme a(dduct) followed by hydrolysis of the low molecu(lar) (non-protein) fraction in 1 N HCl to cleav(e) coenzyme-inhibitor imines, results in relea(se) of PLP and pyridoxal, but not of PMP or (pyri)doxamine (lyases 2 and 4). The acid solub(le) fraction of mixtures of enzyme and cyclo(-) serine reduced with sodium borotritide co(n)tains radioactive fluorescent derivatives w(ith) $\lambda_{max} \sim 340$ nm
Michael reactions	Three types of reagents which impair the activity of eliminating and plurifunctiona(l) lyases (see column C) as a result of nucle(o)philic addition to double bonds (Michael acceptors) fail to inhibit the replacement specific β-lyases (3 and 4, in particular). The reagents are: (a) N-ethylmaleimide; (b) Ala(Cl_2) and Ala(F_3); (c) propargylg(ly) Gly($CH_2-C\equiv CH$)

C	D
Traits of eliminating and plurifunctional lyases	References and comments
Complexes of eliminating lyases (**7, 8**, etc.) with amino substrates and inhibitory quasisubstrates have spectra with conspicuous absorption and negative CD bands at 480–490 nm and in the range above 500 nm	B: our findings (1977) and [10,14] C: Refs. [6–9,20,25] Long-wave maxima ($\lambda \geq 480-500$ nm) are typical of deprotonated E · S- and E · I-ketimines in the tautometric 1,4-quinoid form and of unsaturated $\Delta^{\alpha,\beta}$-substrate-PLP-imines (see [20])
Lyases **5, 6** and **8** have I_{50} values for L-cycloserine (freshly recrystallized) of the order of $10^{-4}-10^{-5}$ M; the affinity for D-isomer is 20–100 fold less. L- and D-cycloserine are bound to the specific protein not by a strong covalent bond (as in transaminases), but rather in the form of imines with $\lambda_{max} \sim 340$ nm; the cycloserine-enzyme complexes are resolved by gel-filtration to (labile) native apo-enzymes and coenzyme-inhibitor imines, yielding mainly PMP and pyridoxamine on mild acidic hydrolysis.	B and C: our data (1977) and [10,14]; γ-cystathionase (**8**), see [25]. Inactivation of PLP-enzymes sensitive to L-cycloserine and its analogs (at least, the transaminases [15–17]) is connected with tautomeric transformation of the coenzyme-inhibitor imine (formation of PMP-ketimine) which entails, in transaminases, acylation of a functional group of the enzyme by decyclized degradation products of cycloserine. According to our data reported earlier [10,14] and in this article, β-replacing lyases do not produce on interaction with L- or D-cycloserine any PMP-ketimines or other stably modified derivatives; they share this feature with glutamate α-decarboxylase [10,14,18] (and possibly other amino acid α-decarboxylases)
a) In E · S complex of eliminating lyases **5, 7, 8**, etc.) N-ethylmaleimide 'traps' LP-$\Delta^{\alpha,\beta}$-imines in reactions $\vec{I}$ and $\vec{III}$ by chiral nucleophilic addition to the β-atom, thus suppressing the formation of α-keto acid. b) Dichloro- and trifluoroalanine covalently inactivate lyase **8** by sequential formation of a coenzyme-quasisubstrate imine, elimination of Hal-H and addition of a nucleophilic group of the protein to the β-C atom in the β-halogenated coenzyme-$\Delta^{\alpha,\beta}$-imine intermediate (activated Michael acceptor [23]).	B, C: data reported in the present article and in Ref. [7,8]; B: (a) [8,20], (b) [23], (c) [22,23]. β-Replacing lyases fail to undergo the mentioned Michael reactions (a, b, c). This constitutes additional evidence supporting our claim that there occurs no intermediate formation of actual unsaturated ($\Delta^{\alpha,\beta}$) coenzyme-substrate imines in their catalytic sites, as opposed to those of other subgroups of the PLP-dependent lyases

Table 4 (continued)

A	B
Criteria	Traits of exclusively β-replacing lyases

a Enzyme numbers (1, 2, 3, etc.) correspond to Table 1.

β-C—X bond and facilitate its fission, replacing the X moiety (without inversion) by the electronegative Y moiety, the leaving group X accepting a proton from the replacing agent or from the aqueous medium.

It has recently been demonstrated (see, for example [8,22,24]) that reactions of nucleophilic addition at the double bond ($\Delta^{\alpha,\beta}$ or $\Delta^{\beta,\gamma}$) of unsaturated ligands formed or bound in the active site, i.e. Michael reactions, which interfere with the normal course of the enzymic process, are a typical property of unsaturated imines formed by eliminating PLP lyases from their amino acid substrates or certain selective inhibitors or inactivators.

As indicated in the comparative listing in Table 4, we have shown the β-replacing lyases to be resistant to a number of agents which affect the eliminating lyases by Michael addition or similar reactions. This further contributes to the evidence in favor of the concept that the reactions catalyzed by β-replacing lyases proceed without the intermediary formation of real α,β-unsaturated (aminoacrylate-PLP) imines. Table 4, summarizing the principal chemical, physical and catalytic properties distinguishing the β-replacement-specific lyases from those of other subgroups, clearly shows the considerable differences in their catalytic mechanisms.

C	D
Traits of eliminating and plurifunctional lyases	References and comments
(c) Propargylglycine by reacting with lyase 8 similarly to the early steps of the normal γ-elimination reaction, irreversibly inactivates the enzyme [22]; the inactivator PLP-imines undergo the following sequential transformations: deprotonation steps in α- and β-sites, attack of the δ-C−H bond in the propargyl sidechain ($^\delta$CH≡C−CH$_2$−) by an acidic group of the protein, resulting in isomerization to an allene group CH$_2$=$^\gamma$C=CH−, which binds the newly formed anionic protein group (a sulfide or phenolate-ion) [22] to its γ-C atom	

Conclusion

The data concisely compared in Table 4 unambiguously demonstrate that α,β-elimination (I) or β-replacement (I + II) reactions catalyzed by eliminating or plurifunctional lyases proceed as shown in Fig. 1, *via* intermediate isomerization of the PLP-substrate aldimines to PMP-ketimines and/or the tautomeric 1,4-quinonoid Schiff bases (2) (see also [20]). This is followed by elimination of an α-H atom and the β-substituent (in the form of an XH molecule) producing an α,β-unsaturated imine (the aminoacrylate-PLP imine, 3), which is then either hydrolyzed to the end products (in elimination reactions, I) or adds an YH molecule to the double bond (Michael reaction), in the case of β-substitution (I + II reactions). (Similar mechanisms for β,γ-elimination (type III) and γ-replacement (type IV) reactions have been discussed by American authors [6−9].)

The sum total of the data presented here also unequivocally indicates that the reactions catalyzed by the β-replacing lyases proceed by an entirely different mechanism. The reaction intermediates include neither PMP-ketimines or their tautomeric 1,4-quinonoid forms, nor $\Delta^{\alpha,\beta}$-unsaturated Schiff base (aminoacrylate-PLP-aldimines). Essential for both the β-replacement reactions as a whole and the stages of α-H labilization and isotopic exchange is the presence of a cosubstrate (replacing agent, YH), appropriate for the given lyase. The size, shape and charge of the YH molecule may vary within broad limits (for example, from H_2O or H_2S to homocysteine for lyases 2 and 3). This excludes the possibility of their functioning as allosteric cofactors in the frame of Koshland's 'induced fit' concept.

The data as a whole indicate that the cosubstrate forms a ternary complex with the PLP enzyme and the amino substrate in which the YH molecule is in proximity to the X–C bond. The molecule can thus contribute to the polarization of this bond, facilitating its rupture in a process coupled with labilization of the α-proton; the bond of the latter to the α-C atom of the amino substrates, while polarized by aldimine formation with the pyridoxal phosphate electrophile, is insufficiently weakened in these lyases. The reaction, therefore, proceeds by a mechanism rather similar to scheme L or Braunstein and Shemyakin [1,2], reproduced in Fig. 2 of the present paper, but with retention of configuration at the C^3 atom; cf. Table 4 and [12].

References

1. Braunstein A.E., Shemyakin M.M. (1952) Doklady of Akad. Nauk SSSR 85, 1115–1119; (1953) Biokhimiya 18, 393–411.
2. Braunstein A.E. (1960) In: P.D. Boyer, H. Lardy, K. Myrbäck, eds.: The Enzymes, Vol. 2. Academic Press, New York, N.Y., pp. 113–184.
3. Metzler D.E., Ikawa M., Snell E.E. (1954) J. Amer. Chem. Soc. 76, 648–652; Snell E.E. (1958) Vitam. Horm. (N.Y.) 16, 77–105.
4. Braunstein A.E. (1953) Usp. Sovrem. Biol. 35, 27–56; (1955) Ukr. Biokhem. Zh. 27, 421–442.
5. Goryachenkova E.V. (1952) Doklady Akad. Nauk SSSR 85, 603–606; 87, 457–460.
6. Snell E.E., Di Mari S.J. (1970) In: P.D. Boyer, ed.: The Enzymes, 3rd Ed., Vol. 2. Academic Press, New York, N.Y., pp. 335–370.
7. Davis L., Metzler D.E. (1972) In: P.D. Boyer, ed.: The Enzymes, 3rd Ed., Vol. 7. Academic Press, New York, N.Y., pp. 33–74.
8. Flavin M., Slaughter C. (1968) J. Biol. Chem. 235, 1112–1118.
9. Dunathan H.C. (1971) Adv. Enzymol. 35, 70–131.
10. Braunstein A.E. (1972) In: J. Drenth et al., eds.: The Enzymes: Structure and Function, FEBS Symposium No. 29. North-Holland Publ. Co., Amsterdam, pp. 135–150; Braunstein A.E. (1974) Izv. Akad. Nauk SSSR, Ser. Biol. (USSR) No. 5, 629–642.
11. Braunstein A.E., Goryachenkova E.V., Tolosa E.A., Willhardt I.H., Yefremova L.L. (1971) Biochim. Biophys. Acta 242, 247–260.
12. Tolosa E.A., Maslova R.N., Goryachenkova E.V., Willhardt I.H., Yefremova L.L. (1975) Eur. J. Biochem. 53, 429–436.
13. Akopyan T.N., Braunstein A.E., Goryachenkova E.V. (1975) Proc. Natl. Acad. Sci. USA 72, 1617–1621.

14. Braunstein A.E., Goryachenkova E.V. (1976) Biochimie (Paris) *58*, 5–17.
15. Karpeisky M.Ya.. Khomutov R.M., Severin E.S., Breusov Yu.N. (1963) In: E. Snell et al., eds.: *Chemical and Biological Aspects of Pyridoxal Catalysis*. Pergamon Press, Oxford, pp. 323–332; Karpeisky M.Ya., Breusov Yu.N. (1965) Biokhimiya *30*, 153–160.
16. Khomutov R.M., Severin E.S., Kovaleva G.K., Gulyaev N.N., Gnuchev N.V., Sashchenko L.P. (1968) In: A.E. Braunstein et al., eds.: *Kimiya i Biologiya Piridoksalevogo Katalisa*. Nauka, Moskva, pp. 381–400.
17. Kovaleva G.K., Severin E.S. (1972) Biokhimiya *37*, 478–484, 1282–1290; Ibid. *37*, 469–477; Severin E.S. (1972) Thesis of Doct. Sci., Moscow.
18. Sashchenko L.P., Severin E.S., Khomutov R.M. (1968) Biokhimia *33*, 142–147.
19. Braunstein A.E. (1973) In: P. Boyer, ed.: *The Enzymes*, 3rd Ed., Vol. 9. Academic Press, New York, N.Y., pp. 379–480.
20. Snell E.E. (1975) Adv. Enzymol. *42*, 287–331.
21. Cleland W. (1970) In: P. Boyer, ed.: *The Enzymes*, 3rd Ed., Vol. 2, Academic Press, New York, N.Y., pp. 1–65.
22. Washtien W., Abeles R. (1977) Biochemistry *16*, 2485–2491.
23. Silverman R., Abeles R. (1976) Biochemistry *15*, 4718–4723.
24. Pestaña A., Sandoval I.V., Sols A. (1971) Arch. Biochem. Biophys. *146*, 373–379.
25. Churchich J.E., Bieler J. (1971) Biochim. Biophys. Acta *229*, 813–815; cf. Brown F.C. et al. (1969) J. Biol. Chem. *244*, 2809–2815.
26. Beeler T., Churchich J.E. (1976) J. Biol. Chem. *251*, 5267–5271.
27. Nomenclature of Multiple Forms of Enzymes (IUPAC-IUB Commission on Biochemical Nomenclature, Recommendations, 1976) (1978) Eur. J. Biochem. *82*, 1–4.

Yu.A. Ovchinnikov and M.N. Kolosov (eds.) Frontiers in Bioorganic Chemistry
and Molecular Biology © 1979, Elsevier/North-Holland Biomedical Press

CHAPTER 10

Total synthesis of a biologically functional gene

H. GOBIND KHORANA

It was my good fortune to come into contact with the late Professor Shemyakin in the fifties when I was working in British Columbia. I still remember with deep sentiment the visit he paid to my laboratory in 1959. Leaving other members of the delegation of the USSR in Eastern Canada, he decided to travel across Canada to talk with my group about nucleotide work. This early interest in nucleic acids and organobiochemistry, in general, showed that among the community of organic chemists, Professor Shemyakin was well ahead of his colleagues. It is a great honour for me to dedicate this article to his memory. Ever since my early contact with Professor Shemyakin, I always found him to be very inspiring.

Although now so well known, the Watson-Crick structure for DNA (Fig. 1) still forms a suitable introduction to my present review. This structure was proposed in 1953 and synthetic work related to this structure immediately began to be my ambition. The important features, from biological standpoint, of the Watson-Crick structure are the two base-pairs, the G-C base-pair, and the A-T base-pair and secondly, the opposite polarity of the two chains. This was really already considered in the original proposal, and it is important in all the work that one does in enzymology or chemistry of polynucleotides. Now, this DNA structure is really 'our way of life' and our religion. Thus, we believe that all mutations which lie at the heart of evolution involve exchange of base-pairs. As mentioned already, it has been really my interest ever since this DNA structure was proposed to try to develop organic chemistry to make DNA, double-stranded and with all the specificity that DNA structure should have. Thus, Fig. 2 focuses more sharply on the chemical features present in each of the two strands of the double-stranded DNA. So, in a hypothetical sequence of a tetranucleotide, we may note the following. There are the four heterocyclic bases, thymine (T), guanine (G), cytosine (C) and adenine (A). The bases are linked to the deoxyribose ring through N-glycosyl bonds and the individual deoxynucleosides are

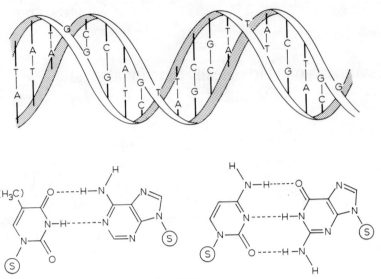

Fig. 1. Watson-Crick structure for DNA.

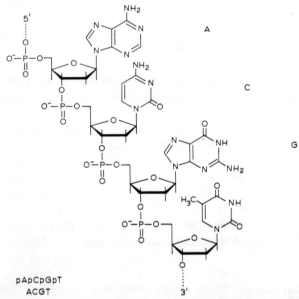

pApCpGpT
ACGT

Fig. 2. A hypothetical tetranucleotide sequence in DNA illustrating chemical details. At the bottom left are shown the accepted abbreviations.

linked to each other through 3′→5′-internucleotide or phosphodiester bonds. Thus, for the synthesis of a specific pre-determined sequence of such chains, we have to learn to connect a nucleotide to the next nucleoside or nucleotide and to work with longer oligonucleotides. I may also point to the conventional abbreviations (Fig. 2) of the tetranucleotide sequence. I would have to use the single letter initials to designate nucleotides for economy of space.

Chemical synthesis of oligonucleotides

We began to learn to make the simplest dinucleotide, namely, thymidine dinucleotide shown in Fig. 3. We took thymidine mononucleotide and used a simple protecting group, the acetyl group, to protect the 3′-hydroxyl function; and in the case of the nucleoside, thymidine, we again used a classical protecting group, a group used by carbohydrate chemists, the bulky triphenylmethyl group. Thus, we obtained the protected nucleoside so that, in principle, we could activate now the phosphate group of the nucleotide and connect it to the 3′-hydroxyl group of the nucleoside. This was done, at that time, luckily, with dicyclohexylcarbodiimide, and the yields were high. Thus in a 1 : 1 ratio of the two components, one obtained, at room temperature, 90–95% yield. So, in this simple case, the situation was highly satisfactory.

I have mentioned the problem of activation of the phosphomonoester group and protection of the primary hydroxyl group. Over the years we have sought to improve the condensing agents. Similarly, except for thymidine, other bases need protection of their amino groups and the trityl group is not satisfactory for purine glycoside bonds. Without going into the systematic development or details of the protecting groups, a summary of the currently used protecting

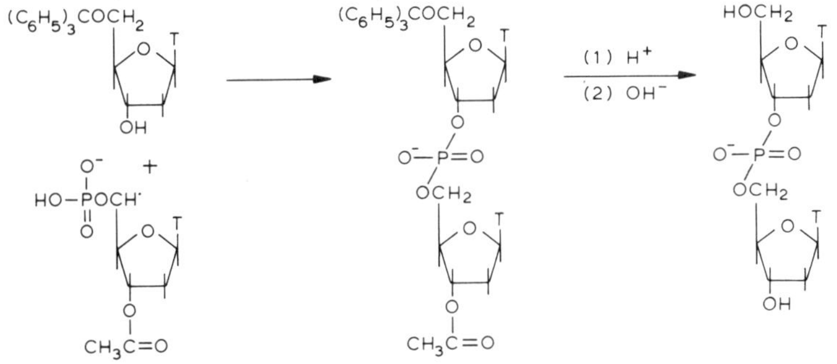

Fig. 3. The synthesis of thymidylyl-(3→5)-thymidine (TpT).

groups as well as of the reagents that we use for activation of the phosphate group is shown in Fig. 4. All of the base-protecting groups are removed by aqueous or methanolic ammonia. As regards the phosphate activation, the mechanism is complex, but the most popular reagent at this time is triisopropyl-benzene sulfonyl chloride. Thus, as seen in Fig. 4, in the nucleotides, we can protect the 5'-hydroxyl by the trityl group that I mentioned, but in the three other cases, particularly in the case of purine glycosyl bonds, they are much more labile than the pyrimidine glycosyl bonds, and we need to use more acid-labile groups. The groups used are monomethoxy- or dimethoxytrityl groups. For the 3'-OH I have already mentioned the acetyl group, but more recently we have introduced substituted silyl groups. For removal of the latter protecting groups simply treatment with fluoride ion is required. And this means that we do not have to give the alkaline treatment at each step. Secondly, the silyl group carrying lipophilic groups gives an enormous advantage in separation by using high pressure liquid chromatography. For the amino groups of the heterocyclic bases,

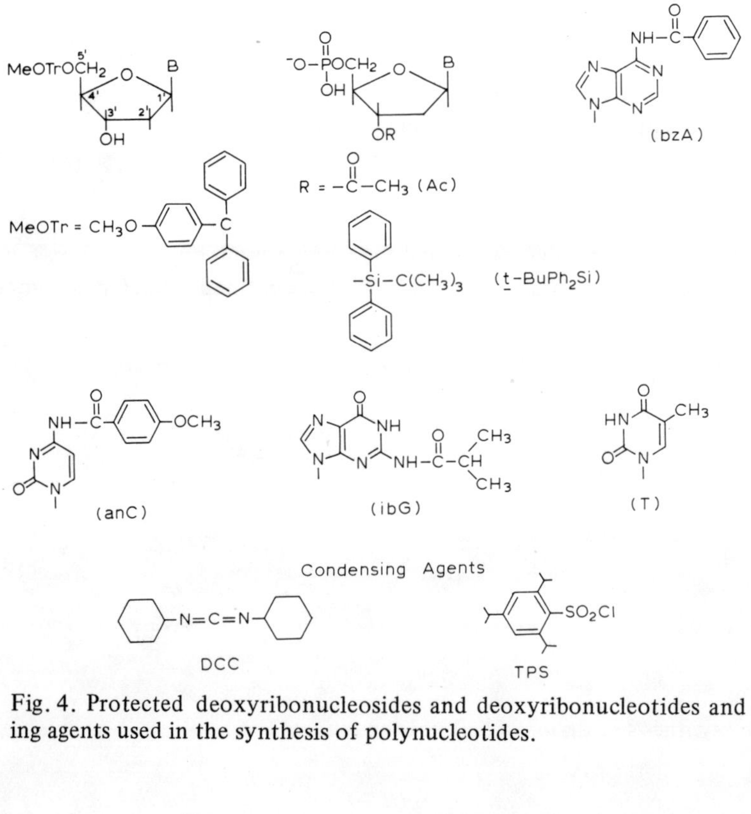

Fig. 4. Protected deoxyribonucleosides and deoxyribonucleotides and condensing agents used in the synthesis of polynucleotides.

the protecting groups are chosen according to the ease of putting these groups on and the ease of their removal by ammoniacal treatment. I should mention in summary that the groups shown are completely satisfactory, are permitted by the DNA chemistry, i.e. no acid treatment is involved, and finally they can all be removed under very mild conditions. For the problem of the activating agents we have, of course, surveyed a whole variety of the condensing agents that have been proposed from time to time. Of the reagents we use, one is the classical dicyclohexylcarbodiimide and the second is the hindered triisopropylbenzene sulfonyl chloride. The mechanisms of these two — it should be noted just in passing — while both complex, differ in important respects.

Fig. 5 shows the steps involved in the synthesis of a trinucleotide. Condensation is first performed between the 3'-OH group of the protected deoxycytidine and the suitably protected thymidine nucleotide. Subsequent alkaline treatment exposes the 3'-OH end group of the dinucleotide, and the condensation with another mononucleotide can then be repeated. The principles are, and must be, analogous to the stepwise and controlled synthesis of macromolecules, polypeptides or polysaccharides, but the problems are enormously complicated by the

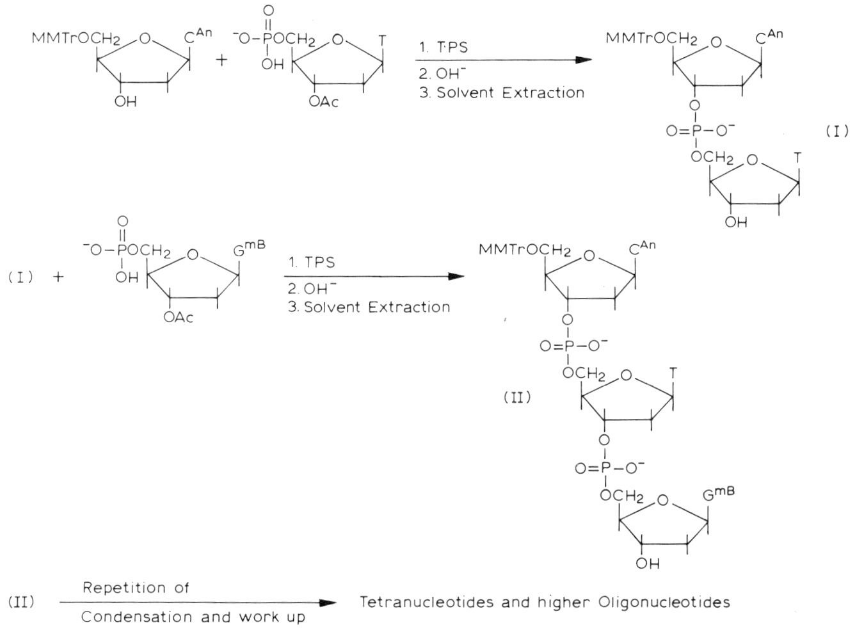

Fig. 5. Principles in the stepwise synthesis of tri- and higher oligonucleotides.

multivalency of phosphoric acid. Fig. 6 gives an extremely simplified outline of the principles involved in the synthesis of a dodecanucleotide. The figure does not convey an impression of the amount of very careful and painstaking work that goes into a synthetic project such as this. In bottom left is shown the desired dodecanucleotide. As is seen, only the one-letter abbreviations are being used for mononucleotides, the diester bonds are simply indicated by hyphens. For the synthesis of the dodecanucleotide of a given sequence, we begin with the nucleoside at the 5'-end, and this is used in the protected form as shown on top in the right half of the figure. Condensations are effected at the 3'-end by bringing in successively protected mononucleotides, or dinucleotides or tri- and tetranucleotides. These are shown in both the protected and the unprotected forms. The actual oligonucleotide blocks that are used are determined by the ease of preparation, economy of chemical synthesis and multiple use of the same block in the synthesis. Thus in the synthesis shown in Fig. 6, intermediates prepared were protected di-, tri-, penta-, hepta-, deca-, and finally, the dodecanucleotide. After each step extensive methods of characterization are used, both at the protected stage and the deprotected stage. The purity of the final products is currently being verified by high liquid chromatography and by the two-dimensional finger-printing methods using radioactive [^{32}P]phosphate group at the terminal 5'-OH group. In this way we have been able to further purify or characterize all the segments that were used in the total synthesis of the gene.

Finally, it should be added that with the development of the hydrophobic silyl group in conjunction with the use of high pressure liquid chromatography

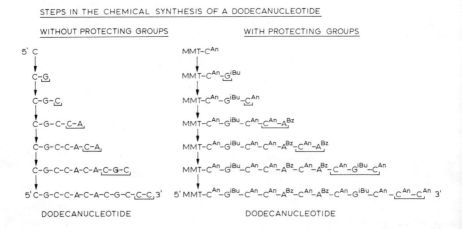

Fig. 6.

for the rapid separation of synthetic products, the time required in chemical synthetic work has been greatly reduced. It may, therefore, be expected that synthetic work in the nucleotide field will proceed with increasing speed.

Principles for the synthesis of double-stranded DNA

The above-described methodology makes it possible to synthesize unambiguously short polynucleotide chains of defined nucleotide sequence. One major application of this methodology was made in the sixties to studies on elucidation of the genetic code. Thus, in conjunction with the use of nucleic acid polymerases, it was possible to amplify and multiply the short sequences prepared by chemical synthesis. The defined messengers thus obtained were used in in vitro protein synthesis. For the next major objective of synthetic work, a decisive discovery was the total primary sequence of a transfer RNA by Holley and coworkers (Fig. 7). The tRNAs are the shortest nucleic acids whose biological functions are clearly defined. There continues to be an enormous amount of interest in the biochemistry, physical chemistry and biology for this group of

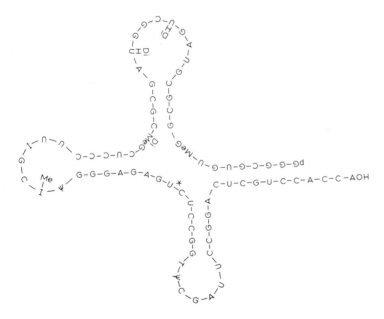

Fig. 7. Clover-leaf model for the secondary structure of yeast alanine tRNA.

RNAs. This is because these molecules are of central importance in the complex machinery of protein synthesis. For example, their key reaction or function is to carry a specific amino acid to the site of polypeptide chain synthesis. The specificity at this step is uniquely important for accuracy in protein synthesis. Further, from chemical and biochemical point of view, we see many intriguing features in these molecules. All of the tRNA's that have been discovered can be written in the four-leaf clover structure (Fig. 7). One always finds common features in regard to the length of the stem, the size of certain loops and the location of certain modified nucleotides. There are always relatively large numbers of modified bases, and, particularly, the base after the anticodon is usually modified. So, because of these many intriguing questions about the structure-function relationships in the tRNA's and, finally, the most important fact that the sequence of the yeast Ala tRNA (Fig. 7) was now known, we began to consider the problem of the synthesis of the DNA corresponding to the entire length of the yeast tRNA.

The question, which had nagged us ever since we could make short chains by chemical synthesis, and was now urgent, was: how is a large DNA of specific sequence to be ever made? Nucleic acids, it has been clear, consisted of very long polynucleotide chains. Chemical synthesis alone would not be able to go very far. As in the study of the genetic code, it was clear that the methods of chemical synthesis had to be coupled to some other concept. And the concept that eventually turned out to be so evident and one that nature used in all of the biology of nucleic acids is, of course, the base complementarity. What we wanted to do is to simply exploit the natural ability of polynucleotide chains to form complementary structures with one another. While work on execution of these ideas was in progress, discoveries of two important enzymes, polynucleotide ligase and polynucleotide kinase were made. In particular, polynucleotide ligase proved to be a boon to our work. This is an enzyme that does really what we want it to do. Biologically, the function of the enzyme seems to be to repair a nick or a break in one of the two strands of DNA. It requires the $3'$-hydroxyl and the $5'$-phosphate groups to be in apposition, and the energy for forming the new diester bond comes from ATP or NAD. Our immediate job was to learn what minimal size double-helical complexes the polynucleotide ligase would require to bring about the joining reactions. Fortunately, our results showed that these chain lengths were quite short, in fact, easily within the range of our chemical synthesis. These studies led to the three-phase methodology for DNA synthesis which is outlined in Fig. 8.

The first phase requires the chemical synthesis of short polynucleotides corresponding to both of the intended strands. A hypothetical sequence is shown in Fig. 8 using the Fisher projection method. In the second step, the $5'$-OH ends

DOUBLE-STRANDED DNA SYNTHESIS

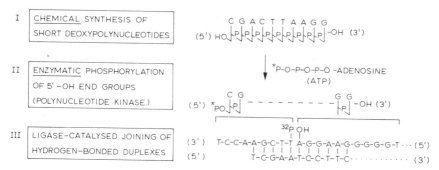

Fig. 8. Three phases in the total synthesis of double-stranded DNA.

of the segments are phosphorylated by using [γ-^{32}P]-labelled ATP. This is done by the use of polynucleotide kinase, another enzyme which has proved to be extremely useful. This enzyme fortunately doesn't worry about the sequence or the length; it simply has the specificity of taking the γ-phosphate of ATP and transferring it, specifically, to 5'-hydroxyl group present at the terminus of a polynucleotide chain. So, all the pieces that we make can be converted in this way to the 5'-labelled oligonucleotides and this, clearly, is important because after having made the large number of required pieces, the main task is join them end to end specifically and accurately. The accuracy can best be monitored by using radioactively labelled phosphate groups. The analyses proper can be done in a variety of ways that I shall mention very shortly. In the third step, three or more segments with overlapping complementary sequences are brought together under suitable conditions of ionic strength and temperature to form double-helical complexes and the ligase is then used to bring about covalent joinings to form continuous duplexes.

Time will not permit a description in detail of the experimental procedures for characterization of the newly formed internucleotide linkages. Suffice it to say that it is possible to be quite certain about this by a variety of methods. For example, one can separate the two strands. There are methods for doing this efficiently, and one can measure the lengths; one can degrade the polynucleotide chain back to mononucleotides to give radioactivity either in the 3' or 5'-nucleotides. Analyses of the radioactivity can establish the participating nucleotides in each enzymatically formed internucleotide bond. With this kind of methodology

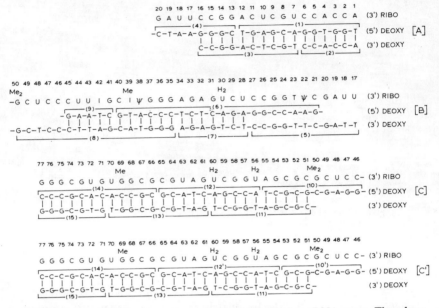

Fig. 9. Total plan for the synthesis of the yeast alanine tRNA gene. The chemically synthesized segments are in brackets, the serial number of the segment being shown inserted in the brackets in parentheses. A total of 17 segments (including 10' and 12') varying in chain length from penta- to icosanucleotides was synthesized.

and going step by step to make duplexes of increasing length, we were able to complete in 1970 the synthesis of the DNA, 77 nucleotides long, which corresponded to the transfer RNA sequence that I have already shown (Fig. 7), the plan of the total synthesis being shown in Fig. 9. Without going into the details of this work, the main point that needs to be made is that with this accomplishment, we had developed confidence in the methodology for synthesizing and characterizing DNA duplexes. Methods had to be developed for separation, isolation and identification and all these basic techniques had been worked out.

The tyrosine tRNA suppressor gene

How did we feel when all this had been done? In terms of DNA chemistry, in terms of being able to say that a biologically specific relatively short DNA can be made in a controlled and step-by-step method in the laboratory, it was satisfying. However, if one is interested in doing interesting biochemical experiments, for example, learning about the mechanisms of start and stop signals in transcription, structure-function relationships in a gene, then from these standpoints, the yeast tRNA gene was a dead end. This is because in terms of protein synthesis, of transcription controls, e.g. promoters, terminators, in terms of enzymology of the enzymes involved in processing tRNA and related problems, the biochemistry of the yeast system was hardly advanced, especially in relation to that of the bacterium E. coli. Thus, from a variety of considerations, we began to undertake the synthesis of a transfer RNA of the bacterium E. coli and the particular gene that we chose was the tyrosine tRNA gene and, in fact, that corresponding to a suppressor tRNA. Governing this specific choice were a large number of reasons at the outset and, fortunately, as the work proceeded genetic and other forthcoming developments further validated the choice of this transfer RNA. One important reason, representing a great advance in transfer RNA field is the discovery of the precursor shown in Fig. 10, a precursor to a tyrosine tRNA, which was discovered in about 1970 by Smith and Altman in Cambridge, England. Now precursors to macromolecules were not new. They had been discovered in the hormone field, in many proteolytic enzymes, in many RNA's; but this was the first time that a precursor to a tRNA had been discovered. Thus,

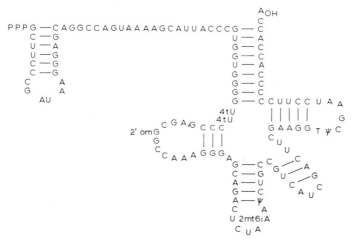

Fig. 10. The primary nucleotide sequence of an E. coli tyrosine tRNA precursor.

what the discovery of this structure clearly showed, was that it had at the 5'-end a triphosphate group on the terminal nucleotide. To those of you who have reflected on the mechanism of copying DNA to RNA by beginning a new chain and using nucleoside triphosphates as the monomers, the following facts become clear. The RNA polymerase in starting a new chain aligns the first triphosphate and the next one, and then a pyrophosphoryl group is eliminated and a diester bond is made; the chain grows in the 5'→3' direction. But the first nucleotide with which the chain begins retains the triphosphate group unless there is a subsequent event — a nuclease action which removes this. For example, in the case of mature transfer RNA's, this indeed is normally the case. Thus a specific cut is made by a nuclease to remove the extra 41 nucleotides at the 5'-end of Fig. 10 to give the stem of a functional transfer RNA. The presence of the 5'-end triphosphate clearly meant that this must be the starting point of the transcription of the gene. In fact, we could begin to have a simple diagram for the different regions and control elements of the gene as shown in Fig. 11. Shown are the regions of the gene (the structural gene) that corresponds to the precursor RNA that can be isolated. This is 126 base-pairs long. The primary RNA product will be cut at the point shown to give the tRNA length. The starting point of transcription must be at the point shown by the arrow and the DNA that presumably precedes it is what we classically call, according to the terminology of Jacob and Monod, the promoter. This promoter is then the region which is somehow recognized by the RNA polymerase. Transcription begins at the indicated point and continues, presumably, until there is a signal that tells it to stop. The latter was the conclusion, or at least our assumption, during most of our work, that the signal for termination may be soon after the C–C–A end; but it turned out as was shown about two years ago, that there actually is no signal for termination. The enzyme continues still beyond the end of this structural gene. However, there is a mechanism for processing, as we found out, soon after the C–C–A end. So, with the introduction of Fig. 11, I would like to define the scope of the things now to be discussed. First, I should briefly review the synthesis of the DNA corresponding to the 126 nucleotide long precursor. We

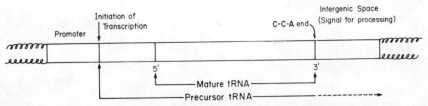

Fig. 11. Diagrammatic representation of the linear arrangement of transcriptional control elements and the structural gene for tyrosine transfer RNA.

know its sequence, we can derive the DNA sequence, and it is a matter of carrying out the painstaking synthesis as well as possible. In further work, the important unknowns are: what are the nucleotide sequences immediately adjacent to the structural gene? Knowing the promoter sequence, we have to define the length that is required for a functional promoter, and finally, to synthesize it. Similarly, in order to probe the region adjacent to the C–C–A end, again the nucleotide sequence must be determined and the aim has to be to include the processing signal into this region. Finally, after the synthesis of the total gene, there should be an enormous scope for doing biochemistry: showing biological activity in vitro as well as by incorporating the gene into a natural system it should be possible to measure the biological activity. So these are really the aspects which the total problem entails.

Synthesis of the DNA corresponding to the precursor (Fig. 10) for tyrosine tRNA suppressor

Fig. 12 shows a total of 26 chemically synthesized segments which together represent the DNA duplex corresponding to the 126 nucleotide long sequence. Note

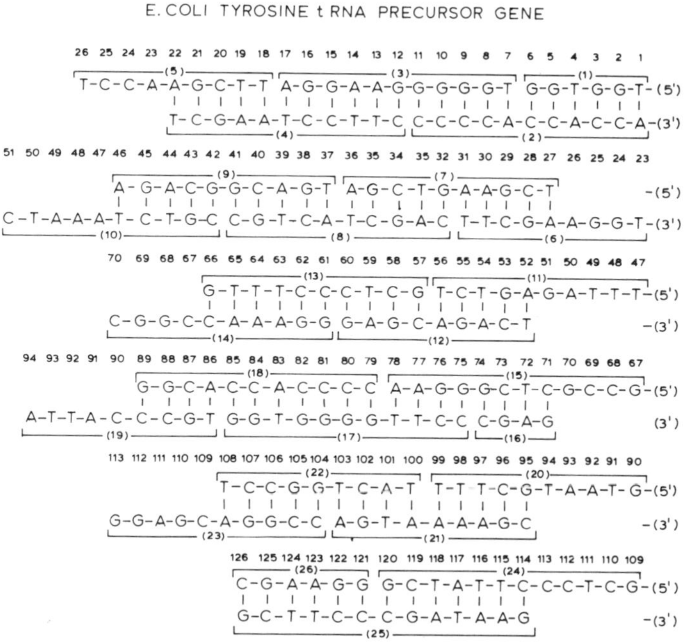

E. COLI TYROSINE t RNA PRECURSOR GENE

```
          26 25 24 23 22 21 20 19 18 17 16 15 14 13 12 11 10  9  8  7  6  5  4  3  2  1
                 ─(5)─              ─────(3)─────          ─────(1)─────
          T-C-C-A-A-G-C-T-T  A-G-G-A-A-G-G-G-G-G-T  G-G-T-G-G-T-(5')
          | | | | | | | | |  | | | | | | | | | | |  | | | | | | |
          T-C-G-A-A-T-C-C-T-T-C  C-C-C-C-A-C-C-A-C-C-A-(3')
                 ─────(4)─────          ─────(2)─────
51 50 49 48 47 46 45 44 43 42 41 40 39 38 37 36 35 34  35 32 31 30 29 28 27 26 25 24 23
                     ─────(9)─────                    ─(7)─
              A-G-A-C-G-G-C-A-G-T  A-G-C-T-G-A-A-G-C-T
              | | | | | | | | | |  | | | | | | | | | |           -(5')
C-T-A-A-A-T-C-T-G-C  C-G-T-C-A-T-C-G-A-C  T-T-C-G-A-A-G-G-T-(3')
        ─(10)─              ─────(8)─────          ─(6)─
              70 69 68 67 66 65 64 63 62 61 60 59 58 57 56 55 54 53 52 51 50 49 48 47
                       ─────(13)─────              ─(11)─
                  G-T-T-T-C-C-C-T-C-G  T-C-T-G-A-G-A-T-T-T-(5')
                  | | | | | | | | | |  | | | | | | | | | |
                  C-G-G-C-C-A-A-A-G-G  G-A-G-C-A-G-A-C-T      -(3')
                       ─(14)─              ─(12)─
94 93 92 91 90 89 88 87 86 85 84 83 82 81 80 79 78 77 76 75 74 73 72 71 70 69 68 67
                       ─(18)─              ─(15)─
              G-G-C-A-A-C-C-A-A-C-C-C-C  A-A-G-G-G-C-T-C-G-C-C-C-G-(5')
              | | | | | | | | | | | | |  | | | | | | | | | | | | |
A-T-T-A-C-C-C-G-T  G-G-T-G-G-G-G-T-T-C-C  C-G-A-G                (3')
        ─(19)─              ─(17)─              ─(16)─
113 112 111 110 109 108 107 106 105 104 103 102 101 100 99 98 97 96 95 94 93 92 91 90
                   ─(22)─              ─(20)─
              T-C-C-G-G-T-C-A-T  T-T-T-C-G-T-A-A-T-G-(5')
              | | | | | | | | |  | | | | | | | | | |
G-G-A-G-C-A-G-G-C-C  A-G-T-A-A-A-A-G-C                    -(3')
        ─(23)─              ─(21)─
                  126 125 124 123 122 121 120 119 118 117 116 115 114 113 112 111 110 109
                      ─(26)─                      ─(24)─
                  C-G-A-A-G-G  G-C-T-A-T-T-C-C-C-T-C-G-(5')
                  | | | | | |  | | | | | | | | | | | |
                  G-C-T-T-C-C-C  G-A-T-A-A-G                -(3')
                              ─(25)─
```

Fig. 12. Plan for the total synthesis of the structural gene for the tyrosine tRNA.

that there will be an overhang or protruding single stranded sequence at the end of each duplex formed after joining reactions. At the outset, in this plan, the first and most important question is: how does one subdivide a DNA as long as 126 nucleotides? There is obviously an enormous number of possibilities. Clearly, jobs such as this should really be done by a satisfactory computer program, a program which should, on the one hand, give proper consideration to the efficiency of the different types of chemical condensations involved and the question of chemical economy. On the other hand, the program should take into consideration the variations in the yields in the ligase reaction. In the latter reactions, we rarely get close to 100% yield; in fact, often we get very much less than 100%. But in the present plan, the first consideration is the presence of suitable overlaps for the enzyme reactions. Secondly, segments with possibilities of self-complementary structures must be avoided. Self-structures are easily possible because DNA has only two base-pairs and in a long DNA, there could be, on the average, 50–60% possibility for self-complementarity (for a chain to fold back on itself or making hairpins when given single strands). The next important con-

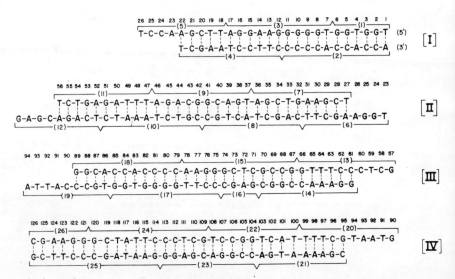

Fig. 13. Plan for the total synthesis of the gene for the precursor of tyrosine tRNA. Subdivision of the total chemically synthesized segments into four parts for the purpose of enzymatic joining.

sideration is chemical economy; in other words, the multiple use of the blocks or chemical segments synthesized. Thus looking at the total sequence, there are nonanucleotides, heptanucleotides, hexanucleotides etc., which occur more than once. From these considerations, we finalized the plan shown in Fig. 12. It took several years to complete the total synthetic job, but, of course, it would take much less time now, because of the more rapid separation methods.

Having made all the segments, the next question is: where does one start with the enzymatic joinings? Joining still requires actually a large amount of empirical

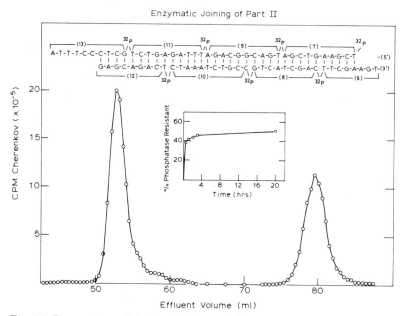

Fig. 14. Preparation of duplex [II] containing segments 6 to 13. Reactions containing 5'-^{32}P-labeled segments 6, 7, 8, 9, 10, 11, and 13 were used in 5'-phosphorylated (^{32}P) form while segment 12 was unphosphorylated. The kinetics of the formation of phosphatase-resistant radioactivity are shown in the inset. The reaction mixture contained 100 mM Tris, pH 7.6, 10 mM dithiothreitol, 10 mM MgCl$_2$, 500 µM ATP, and 40 µM concentration of each one of the oligonucleotides except for segment 6 which was present at 60 µM. The segments were annealed by slow-cooling from 95° to 5° over a 4-hour period and the ligase was added to a concentration of 400 units/ml. After 20 hours, the reaction mixture was passed through a column (1 × 150 cm) of Bio-gel A-0.5 m (200 to 400 mesh). The eluant was 50 mM triethylammonium bicarbonate. The first peak contained the joined product and it was characterized. The second peak contained the unreacted starting materials.

work. After a great deal of preliminary surveys, we were able to form four subgroups out of the total 26 pieces, and these are shown in Fig. 13. Thus, the top duplex (duplex I) consists of five segments while duplexes II–IV contain 6–7 segments. Without going into the details of the enzymatic reactions which have all been published, I will choose one example. This is the synthesis of duplex II. In fact at the time we did this, we used 8 segments, segments 6 to 13 (Fig. 12). The case chosen is particularly satisfying, in that it is possible to go from chemically made short pieces, in one step, to a DNA of 44 nucleotide basepairs. The joining reaction and the isolation of the DNA are shown in Fig. 14.

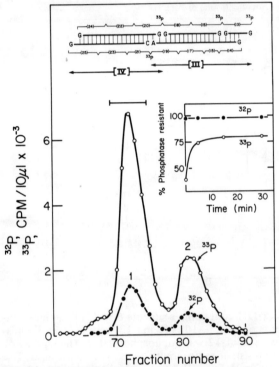

Fig. 15. Joining of duplex [III] (segments 13 to 19) to duplex [IV] (segments 20 to 25). The duplexes as prepared in accompanying papers (190 pmol of each) were annealed at 37° in the presence of the standard components (Tris buffer, pH 8.0, 50 mM and $MgCl_2$, 10 mM). After cooling to 4°, ATP (0.1 mM), dithiothreitol (2 mM), and polynucleotide ligase (60 units/ml) were added, the total volume being 100 μl. The joining was followed by the phosphatase assay (see inset) and after 30 min, the reaction was stopped with EDTA and the mixture separated by flow through an Agarose 0.5 m column as shown.

The segments can be seen in the aligned duplex shown in the inset. They all have ^{32}P's at the 5'-ends, the polarity of the two strands is opposite. The kinetics of joining are shown in the inset. The joinings are measured by the formation of the phosphatase-resistant radioactivity. The separation is done on a sieving (Sephadex, Bio-gel etc.) column. On the left, emerging first, is the fully joined product which can be characterized to be correct in all respects. Unreacted components emerge later (second peak). We still don't understand why we don't get 100% reactions in these enzymic joinings. That is the only example I had planned to give to illustrate joining of single-stranded pieces to form duplexes. In this way, the four duplexes shown in Fig. 13 were made and fully characterized. The last

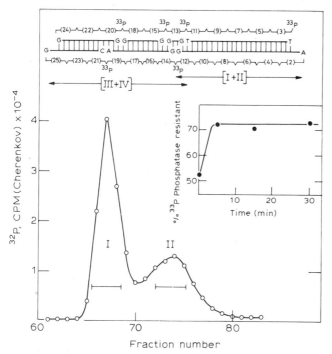

Fig. 16. The joining of duplex [I + II] to duplex [III + IV]: synthesis of the total duplex [I + II + III + IV]. The reaction mixture set up exactly as in the experiment of Fig. 15, contained in a reaction volume 100 μl, the following components: duplex [I + II] (100 pmol), duplex [III + IV] (100 pmol), Tris, pH 8 (50 mM), MgCl$_2$ (10 mM) ATP (0.1 mM), dithiothreitol (1 mM), and ligase (400 units/ml). The reaction was performed at 4° and about 0.5 μl or less aliquots were removed for phosphatase sensitivity as analyzed by DEAE-cellulose paper assay. The kinetics are shown in the inset. Separation by flow through an Agarose column is shown.

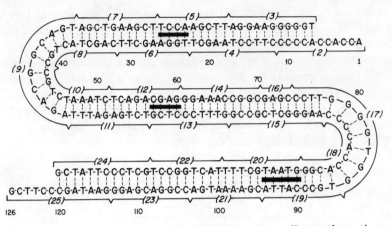

Fig. 17. The totally synthetic DNA duplex corresponding to the entire sequence (126 nucleotides) of the precursor for an E. coli tyrosine suppressor tRNA. The numbers and distances between the *carets* show the oligonucleotides which were synthesized chemically. The *carets* indicate the sites where joining was accomplished by the use of polynucleotide ligase. The *three thick strips* between the duplex indicate the sites where (a) duplex [I] and duplex [II] (between nucleotides 23 and 26), (b) duplex [III] and duplex [IV] (between nucleotides 90 and 94) and (c) duplex [I + II] and duplex [III + IV] (between nucleotides 57 and 60) were joined.

phase is the joining of the short duplexes through their protruding single-stranded ends (Fig. 13) to form the total duplex. Fortunately, joinings at this stage are very efficient and rapid. I will show two examples to illustrate the final phase of this work. Fig. 15 shows the joining of duplex III to duplex IV, in this case both, [^{32}P] and [^{33}P] being used to selectively label the ends. Thus the joinings are extremely rapid, being complete within a few minutes and satisfactory. Extensive analyses already published show accuracy of joinings. Similarly, Fig. 16 shows the joining of the two-halfs of the precursor DNA and separation of the total DNA. Fig. 17 is a summary structure which shows the synthetic steps used and the DNA products finally obtained.

Sequences and synthesis in the promoter region and the region adjoining the C–C–A end of tyrosine tRNA suppressor gene

The synthetic work brings us to the regions which somehow control the start of transcription and either the termination of transcription or the processing of the primary transcript. The next job is to do the sequences to the left and to the right of the structural gene which we have synthesized. Now it is about ten years ago that this question was posed. At that time, doing DNA sequence seemed to be an impossible job, but we decided to make a start. Since then DNA sequencing has undergone a revolution, and sequences are coming out in lengths of hundreds and even thousands of nucleotides, but at that time it was not so. The first and most important question was: how to focus specifically on the short regions at the ends of the gene? Consider the DNA of E. coli; it is of about three billion mol. wt. Out of this total length, we are interested only in about $20-50 \times 10^3$ mol. wt. fragment at a specific place. One approach to focus on a specific region of DNA is to use a specific protein which will selectively recognize and tightly bind to the region of interest. This is the approach which was used by Gilbert as well as by Schaller and colleagues and others for the isolation of the operator region in the lactose operon and a promoter, respectively. When the sequence adjoining the desired region is known, as is the case for the transfer RNA gene, then another approach is possible. In this, one may separate the two strands of the DNA and specifically hybridize a shorter polydeoxynucleotide at the adjacent site where the sequence is known. A primer-template relationship is thus established, and if the polarity of the primer piece is correct, then the 3'-end of the primer may be extended by means of DNA polymerases. The latter would bring about nucleotide incorporations according to the nucleotide sequence in the template strand. The nucleotide sequence of the unknown region can then be deduced from the pattern of nucleotide incorporation. In the case of the tyrosine tRNA gene, it is fortunate, as mentioned above, that one tRNA gene, called the suppressor gene, can be inserted into the DNA of the temperate bacteriophage $\phi 80$. The resulting bacteriophage carrying the suppressor tRNA gene is called $\phi 80 psu_{III}$. The phage can be readily multiplied in quantity, and therefore offers a very convenient source for sequence determination of the tRNA gene. The scheme used for determination of the nucleotide sequences at the two ends of the structural gene is shown in Fig. 18. Thus, it is clear that the elongation of the piece hybridized at the C–C–A end and of the r-strand of $\phi 80 psu_{III}$ takes us into the terminator region, whereas hybridization of a piece complementary to the l-strand can take us into the promoter region.

The three requirements for a start on the sequence work by this approach are: first, effective separation of the two strands of the DNA; second, synthesis of the adequately long synthetic polynucleotide to serve as a primer, and third, the specific hybridization at the appropriate site of the very long DNA so that one can be certain that the nucleotide addition occurs strictly at the desired region. All these conditions were satisfactorily met.

Without going into details of methods used in determination of the sequences,

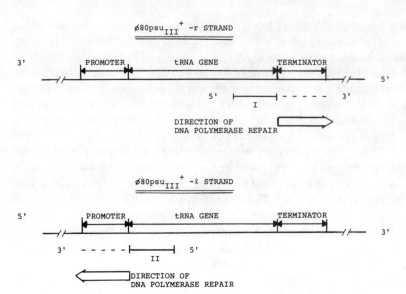

Fig. 18. Schematic presentation of the plan for sequencing of the terminator and the promoter regions of the tyrosine tRNA gene using primer-template complexes.

Fig. 19 is presented to show the principles only. Thus, for determining the promoter sequence, the primers DNA-I and preferably DNA-II are hybridized to the l-strand of $\phi80psu_{III}^+$ DNA. In the arrangement of Fig. 19 the starting point of transcription and the start of the promoter region are shown. DNA-II when extended by five nucleotides will be at the start of the promoter region. As illustrations of the types of devices used to keep the newly growing chains within manageable size, the following points may be noted. In the first step (A in Fig. 19), three deoxynucleoside triphosphates were provided in the polymerase reaction mixture. The absence of dTTP limited the chain growth to the sequence shown in the dotted box in A. The chain thus elongated was used again as a primer with a different set of three triphosphates (B in Fig. 19). A third device used for selective fragmentation of the newly grown chain was the substitution of rCTP in place of dCTP. This conferred alkaline lability at specific sites in the chain. Finally, in experiment C of Fig. 19, the general concept was to add three nucleoside triphosphates at standard concentrations but the fourth triphosphate (in this case, dTTP) in a rate-limiting amount. In this way successively longer chains useful in sequencing were obtained. This work was carried out several years ago and it is of interest to note that these concepts are widely used in the current methodology for DNA sequencing on a much greater scale.

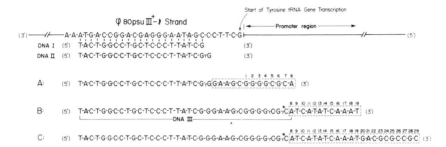

Fig. 19. Experimental plan for sequencing and the nucleotide sequences determined in the promoter region of the tyrosine tRNA gene. The primer-template complexes were initially obtained by hybridizing DNA-I — DNA-III to the l-strand of the φ80psu$_{III}$+ DNA. DNA polymerase-catalyzed elongations were carried out using three nucleoside triphosphates at a time. The new nucleotide sequence discovered after each elongation and subsequent alkaline cleavage and analysis is shown in the appropriate dashed box.

The total promoter sequence as far as determined is shown in Fig. 20 — A. The arrow at the top right shows the direction of transcription, nucleotide no. 1 indicating the starting point. The promoter sequence determined is from nucleotide −1 to −59 and is shown in the double-stranded form. The sequence possesses interesting elements of symmetry that are indicated by dashed and solid boxes and their relationship to each other by the direction of the arrows. Unfortunately, the significance of the symmetry in the sequence remains unclear, especially since its presence or extent is variable in the different promoters that have been sequenced. How did we know that we had covered in our sequence the length required for a functional promoter? The answer simply is the cumulative evidence which had been forthcoming from different laboratories and the work that we ourselves had carried out. It seemed that we did not have to go further than about 50 or so nucleotides or perhaps the minimal length was between 40 and 50 nucleotides. Therefore, in Fig. 20 — B, we show the decisions that we made for the synthetic purpose. Thus, the synthetic plan had some modifications. From −1 we decided to go up to −51 with the natural sequence, since we thought that this would be sufficient. But then we did one more thing at this left end: we added a piece which is a specific recognition sequence for one of the DNA splicing enzymes, the EcoR$_1$ restriction enzyme. This was done, and a similar measure was taken at the other terminus of the gene, in order to integrate our gene eventually into a vector and be able to measure its biological activity. The terminal hexanucleotide sequence shown is the specific sequence of

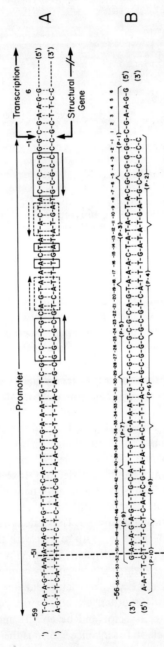

Fig. 20. (A) The nucleotide sequence of the promoter of the tyrosine suppressor tRNA gene. The point of initiation of transcription and its direction are shown. Elements of two-fold symmetry in the sequence are shown in the boxes. (B) Plan for the total synthesis of the promotor region of the tyrosine suppressor tRNA gene. Included are the 51 nucleotide base-pairs in the promoter region + one C : G base-pair + the single stranded A–A–T–T sequence at the 5′-end. Together the A–A–T–T–C sequence serves as the recognition sequence for the EcoR$_1$ restriction enzyme. The ten segments to be synthesized are indicated by horizontal brackets.

an enzyme that comes from E. coli. Now, to briefly review the synthetic plan (Fig. 20 — B), we have divided the total length into 10 pieces, and Fig. 21 briefly shows the steps used in the total joining because of the complications caused by the elements of symmetry. The plan in Fig. 21 ultimately got around the difficulties. Thus, as shown, we had to do the joinings in four steps: Step 1 meant

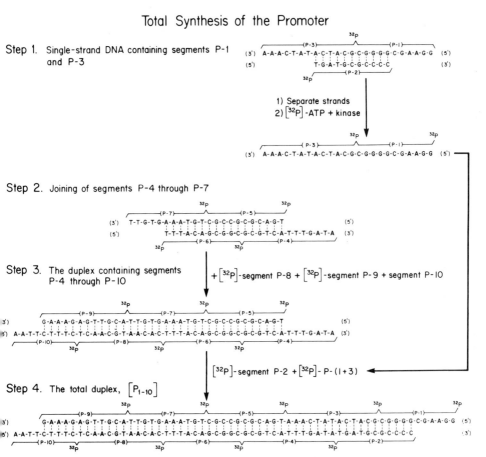

Fig. 21. Plan adopted for the stepwise joining of the chemically synthesized segments P–1 to P–10 to form the total DNA corresponding to the promoter. Thus, in Step 1, segments P–1 and P–3 were joined. In Step 2, segments P–4 to P–7 were joined to form the corresponding duplex. In Step 3, the duplex from Step 2 was extended by joining in segments P–8 to P–10. Finally (Step 4), segments P–(1 + 3) and P–2 were joined to the duplex [P_{4-10}].

joining the segment P–1 to P–3 with P–2 as template; the next step involved joining segments P–4 to P–7, then elongating this to the duplex containing P–4 to P–10 (Step 3). Finally, in Step 4, the total synthesis was completed.

This completes a brief review of the promoter sequence and its synthesis. In the above, the total synthesis of the structural gene has already been completed. Now, attention is turned to the part of the gene beyond the C–C–A end of the tRNA. The main problem at this end again was – how much sequencing will have to be done to cover the signal for the processing of the gene transcript, because the signal for termination of transcription was a very long way off. The sequence that we found at the opposite end (the amino acid acceptor end) is shown in Fig. 22 in the double-stranded form. We went only 23 nucleotides beyond the terminus of the tRNA. And here again we found a palindrome, and the question was as to its significance. There were speculations about its significance for termination of transcription. What turned out was extremely interesting in a different way. Thus, it has become evident that genes may not be transcribed individually but in clusters. The primary continuous RNA product may then undergo specific cuts and further processing to give functional RNA's. This certainly is the case with our gene, and the role of the palindrome in top half of Fig. 22 turned out to be that its transcript is able to form a hairpin which is specifically recognized by an endonuclease. The resulting RNA is further digested away step by step to the point of C–C–A sequence. This discovery was, of course, made possible by synthesis of the duplex shown in Fig. 22, its joining to the full length of the synthetic structural gene, and subsequent experiments

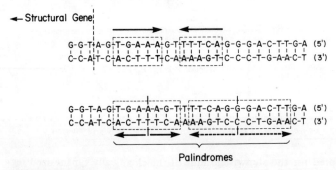

Fig. 22. Elements of symmetry in the nucleotide sequence beyond the 3′-end of the tyrosine tRNA gene. At the top, the double-stranded configuration of this region is shown. The symmetric sections about an axis perpendicular to the page are indicated in the dashed boxes. The single strand region of the template r-strand is shown at the bottom. There are two symmetric regions as indicated.

on transcription and processing of the transcription. Fortunately, as a result, we did not have to go further with sequencing or synthesis in this direction. The duplex Va in Fig. 23 shows the manner this part of the gene was synthesized. Indeed, we were able to further modify this part to the duplex Vb. It now con-

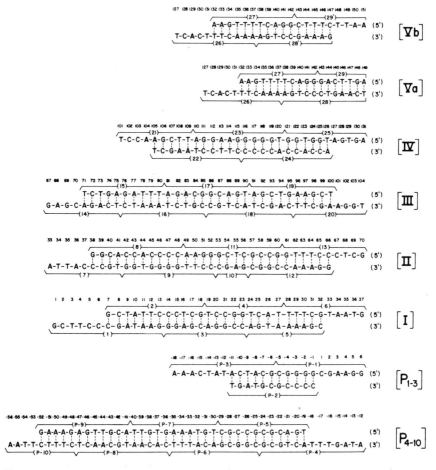

Fig. 23. The synthetic duplexes which comprise the promoter [$P_{(1-3)}$ and $P_{(4-10)}$], the structural gene corresponding to the Altman-Smith tyr tRNA precursor (duplexes [I] – [IV]) and to the region adjacent to the C–C–A end of the tRNA [Va], [Vb]) which contains the signal for processing of the primary transcript in this region of the tyrosine tRNA. Duplex [Vb] contains, in addition, the $EcoR_1$ endonuclease-specific sequence at the appropriate terminus.

tained only a short naturally occurring sequence of 16 nucleotides adjoining the C–C–A end, and the terminal artificial sequence, also used at the promoter end, now contained the EcoR₁ restriction enzyme sequence. In this way, both termini of the synthetic gene would carry the protruding sequence for insertion to suitable vectors.

Total synthesis of the tyr tRNA gene

Fig. 23, in fact, contains the total duplexes, eight in all, resulting from 40 chemically synthesized segments. Shown are the two duplexes P_{1-3} and P_{4-10} (at bottom) belonging to the promoter region, duplexes I–IV belonging to the

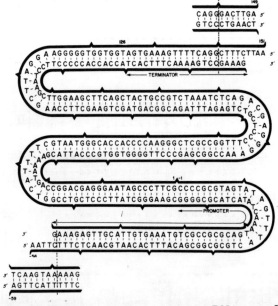

Fig. 24. Synthetic tyrosine suppressor tRNA gene. From bottom left, following the terminal EcoR₁, endonuclease-specific sequence, the DNA contains a 51 nucleotide-long promoter region, a 126 nucleotide-long DNA corresponding to the precursor RNA discovered by Altman and Smith and finally a 25 nucleotide-long DNA. Of the latter, 16 nucleotides belong to the natural sequence adjoining the C–C–A end and the remainder is a modified sequence including the EcoR₁ endonuclease-specific sequence. Shown on top right and bottom left are the unaltered sequences as they occur naturally at these sites.

structural gene and duplexes Va and Vb to the region beyond the C–C–A end. Without going further into details of joining reactions, we now present in Fig. 24 the totally synthetic gene. The additional sequences on top right and at bottom left are the natural sequences as they continue in the two directions.

Transcription in vitro of the synthetic gene and related DNA's

Throughout these studies, the prominent goals undoubtedly were (1) the in vitro transcription of the synthetic gene followed by all the biochemistry involved in the formation of a mature functional tRNA, and (2) demonstration of the biological activity, namely suppression of amber mutation, in vivo of the synthetic gene. The next two Figs. (Figs. 25 and 26) are devoted to a very brief summary of the transcription work. Studies on transcription of short synthetic single-stranded and double-stranded DNA's had been undertaken already in our laboratory in the late sixties, concomitant with the work on the synthesis of the genes. By using ribo-oligonucleotide primers, for example, DNA-I (Fig. 25) had been transcribed quite precisely using carefully controlled conditions. Next, when DNA II, which consisted of the structural gene, the duplex beyond the C–C–A region and a short oligonucleotide at the promoter end, became available, attempts were made to transcribe the DNA in a strand-specific fashion with the help of the ribotetranucleotide, C–C–C–G, primer (Fig. 25). (Work on the promoter was not complete at this stage.)

The results were very gratifying and the primary transcript was shown to consist, largely, of the RNA shown in Fig. 26. The various important features are indicated. The two arrows show the cuts made by the two distinct endoribonucleases. The nucleases acting on the stem of the tRNA had been described previously, but it was most exciting to see a cut in the hairpin at the right end, caused by the presence of the palindrome in DNA of this region, made by a second hitherto undescribed endonuclease. An exonuclease then removed the extra seven nucleotides to give the C–C–A terminus. Other base modifications are also performed by the enzymes present in E. coli extracts. These products have been thoroughly characterized.

Biological activity of the synthetic gene

Suppression of amber (non-sense) mutation is the second important experiment to be performed with the synthetic gene. Again, the experiment would be briefly reviewed, omitting the extensive plasmid work carried out. Now, we had chosen

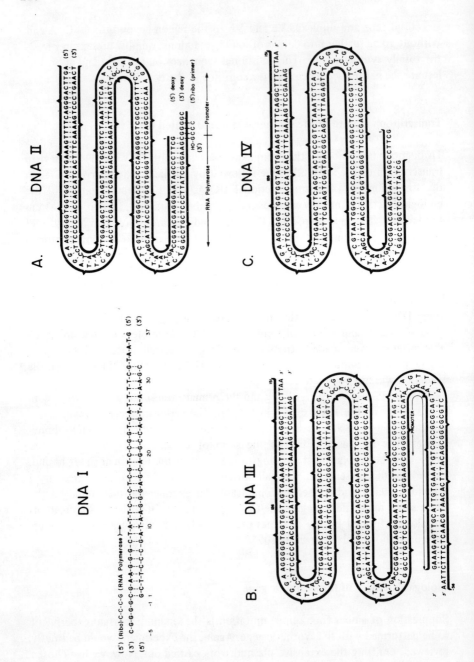

Fig. 25. Different DNA's used in the study of transcription. DNA I: this contains the duplex [I], which covers the sequence 1-37 of the structural gene for tyrosine suppressor tRNA, plus the promoter segment P–1. The DNA together with the ribo-tetranucleotide primer, r–C–C–C–G, was used in initial studies of primer-dependent transcription. DNA II, which contains DNA I at one end plus duplexes [II], [III], [IV] and [Va], was used for primer-dependent transcription to form the tRNA precursor and to study its processing. The tetranucleotide primer and direction of transcription are indicated. DNA III is the synthetic gene which, in contrast with DNA II, contains 56 nucleotides in the promoter region. Out of these, 51 base-pairs are natural promoter sequence and the remainder correspond to the EcoR$_1$ sequence. In the region following the C–C–A end, this DNA contains only 16 base-pairs of the natural sequence while the remaining sequence of 10 nucleotides is artificial with EcoR$_1$ endonuclease sequence at the 5'-end. DNA IV differs from DNA III in lacking the promoter region. DNA IV was used as a control in promoter-dependent transcription experiments.

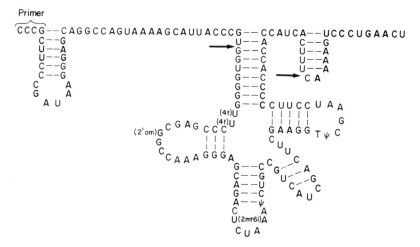

Fig. 26. Sites of endonucleolytic cleavages in processing of the primer-dependent transcript of DNA II (Fig. 25). The arrow on the left is the previously determined site of cleavage by endonuclease P. The arrow on the right is the site now determined as the first step in processing of the 3'-end of the transcript.

a transfer RNA that would bring about what biologists in the late fifties and early sixties had discovered, namely, biological suppression. I should explain what suppression is very briefly. Let us take a bacteriophage that infects E. coli for example, T4 phage. When it infects, it takes over the machinery of the cell; it multiplies and the cell eventually dies and a large number of copies of mature phage particles are produced. In any case, the important point is that all the functions that the phage chromosome itself brings in for successful infection are

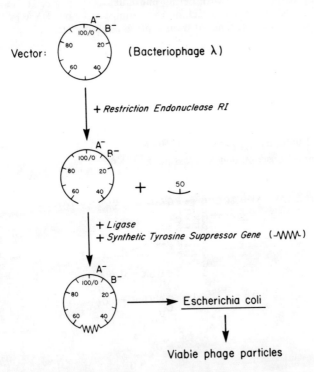

Fig. 27. Cloning of the synthetic gene for the tyrosine suppressor tRNA. The vector used was a derivative of bacteriophage λ with two amber mutations (A^-, B^-). Digestion with restriction endonuclease R_1 excised a non-essential piece (at 50 min of genetic map) out of the total genome. Subsequent addition of the synthetic gene (—∿∿∿—) and ligase gave the circular phage containing the synthetic gene. This on infection of E. coli produced viable phage particles. The results are shown in Fig. 28.

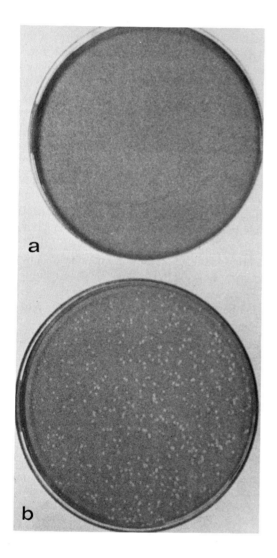

Fig. 28. Experiment showing phage growth in E. coli following insertion of the synthetic amber suppressor gene. The two culture dishes show (a) bacteriophage λ with amber mutation and (b) the same vector with inserted suppressor tRNA gene, prepared as in Fig. 27. The phage colonies were present only in (b).

obviously vital to its replication. It can happen that a mutation occurs in one of the vital genes of the phage, such that the phage cannot grown and multiply in the cell. Thus, the infection fails. Now another biological event can occur such that E. coli can also develop a mutation elsewhere which will undo the harm of the first mutation. In other words, a separate mutation somewhere else, and it was shown later to occur in the transfer RNA, would overcome the harmful effect of the first mutation. This is called biological suppression and it works at the level of protein synthesis. Now, if we look back on the tRNA structure shown in Fig. 10, the reason why we chose to work with the gene of this structure is as follows. The anticodon in this transfer RNA contains a mutation such that it has mutated from the normal tyrosine anticodon, GUA, to CUA. The latter can now recognize the nonsense codon UAG. Thus antiparallel base-pairing occurs satisfactorily and so this tRNA, because it can read or recognize this amber or nonsense mutation enables the phage to grow in E. coli. So, the test for the suppressor activity that is done is the following. We take a phage which has such a mutation and therefore cannot grow in E. coli. We will put our gene into E. coli through a vector. The E. coli carrying our synthetic gene when infected with the mutant T4 phage would be able to bring about suppression and the phage should be able to grow. The way this is done is shown in Fig. 27 and 28. Shown in Fig. 27 is a vector (bacteriophage λ). We take out a piece which is not necessary for the cell by using the restriction endonuclease R_1. Addition of synthetic gene, which has the R_1 endonuclease sequences at the termini, and the ligase, we obtain certain circular molecules with the gene properly inserted as shown. Now we want to compare the results when E coli is infected with the starting vector (Fig. 28a), which contains two amber mutations (A^-, B^-), and the vector carrying the synthetic gene (Fig. 28b). It is clear from the sample of results shown in Fig. 28 that only the vector with the suppressor gene (Fig. 28b) was able to grow and form phage colonies. At least three different cloning vectors or plasmids have been used, all these confirming the biological suppression brought about by the synthetic gene.

Concluding remarks

The 'distant' hope of the synthesis of a biologically functional nucleic acid was expressed some twenty years ago [8]. Successful completion of the total synthesis of a gene carrying its own promoter as well as the signals for the processing of the primary transcript to a functional suppressor tRNA has now been accomplished. The synthesis of 207 base-pair long DNA entailed the formation of about 400 internucleotide bonds by chemical condensation and the subsequent enzymatic joining of 39 oligonucleotides. This and the earlier work from this laboratory demonstrate that reasonably satisfactory methodology exists for the step by step, controlled and unambiguous synthesis of macromolecular DNA. In brief review, total synthesis, requiring no preformed template strand, consists of three phases: (1) the chemical synthesis of short oligonucleotides encompassing both strands; (2) the joining of the single-stranded oligonucleotide, several at a time, to form relatively short duplexes with, usually, 5'-OH termini protruding (the latter would facilitate the end phosphorylation by the polynucleotide kinase); and (3) the final joining of the short duplexes, whenever desired, to form the total DNA. It has been previously emphasized that the slowest and most laborious phase is the chemical synthesis and purification of the oligonucleotides. Recent developments in this laboratory using high pressure liquid chromatography as well as the introduction of the novel highly lipophilic protecting groups have contributed greatly to the shortening of time required in chemical synthesis.

The way is now open for a systematic attack, by using the present approach, on a number of major problems of biological interest. Firstly, studies on the structure-function of the tRNA molecule may begin by performing designed base-changes in the critical regions of the DNA specifying the tRNA. Similarly, further studies on the mechanisms of base-modifications and of the enzymes processing the primary transcripts are now possible. Finally, studies on the structure-function relationships in the promoter region can now be undertaken. Despite the fact that a large number of promoters has been sequenced, the structural features important in (a) the recognition of the promoter region by the RNA polymerase, (b), the formation of the stable complex and (c), the point of initiation of transcription, remain poorly understood.

Studies on the structure-function relationships of the genetic control elements and the specific proteins recognizing them are now increasing rapidly and, hopefully, the synthetic methodology now described will facilitate such studies. Until the recent dramatic progress in the determination of DNA sequences, the degrees of freedom in applying bio-organic synthesis to such problems were too great. However, now with the frame work for studies of protein-nucleic acid interactions attaining more precision, synthesis can provide systematic approaches to the precise studies of the control regions on the genomes.

Finally, in the general field of recombinant DNA, possibilities with enormous scope are opening up in which the present methodology may complement the genetic approaches. For example, it may not be necessary to synthesize full lengths of the genes with their own control elements as was done in the present work. Instead, shorter, more easily manageable segments corresponding to the necessary gene products, e.g. peptide hormones, may be synthesized and inserted into the plasmids or vectors.

Selected references

For literature citations and details of work, on which this lecture is based, the following selected papers from the author's laboratory may be consulted.

1. Khorana H.G., Agarwal K.L., Büchi H., Caruthers M.H., Gupta N., Kleppe K., Kumer A., Ohtsuka E., RajBhandary U.L., Van De Sande J.H., Sgaramella V., Terao T., Weber H., Yamada T. (1972) Studies on polynucleotides. CIII. The total synthesis of the structural gene for alanine transfer ribonucleic acid from yeast (General Introduction). J. Mol. Biol. *72*, 209, and the accompanying papers.
2. Khorana H.G. (1974) The structure and synthesis of a tyrosine transfer RNA gene, Lecture at XXIVth IUPAC Congress, Hamburg, August 1973. Pure Appl. Chem. *2*, 19–43.
3. Sekiya Takao, Khorana H. Gobind (1974) Nucleotide sequence in the promoter region of the E. coli tyrosine tRNA gene. Proc. Natl. Acad. Sci. USA *71*, 2978.
4. Khorana H. Gobind (1975) A bacterial gene for tyrosine transfer RNA: promoter and terminator sequences and progress in total synthesis. In: Mano E.B., ed.: *Proceedings of the International Symposium on Macromolecules,* Rio de Janeiro, 1974. Elsevier Scientific Publishing, Amsterdam, pp. 371–395.
5. Khorana H. Gobind, Agarwal Kan L., Besmer P., Büchi H., Caruthers M.H., Cashion P.J., Fridkin M., Jay E., Kleppe K., Kleppe R., Kumar A., Loewen P.C., Miller R.C., Minamoto K., Panet A., RajBhandary U.L., Ramamoorthy B., Sekiya T., Takeya T., Van De Sande J.H. (1976) Studies on polynucleotides. CXXXI. Total synthesis of the structural gene for the precursor of a tyrosine suppressor transfer RNA from E. coli (1). General introduction. J. Biol. Chem. *251*, 565, and the accompanying papers.
6. Sekiya Takao, Gait M.J., Norris K., Ramamoorthy B., Khorana H. Gobind (1976) The nucleotide sequence in the promoter region of the gene for an E. coli tyrosine transfer ribonucleic acid. J. Biol. Chem. *251*, 4481.
7. Sekiya Takao, Contreras R., Küpper H., Landy A., Khorana H. Gobind (1976) Escherichia coli tyrosine transfer ribonucleic acid genes: nucleotide sequence of their promoters and of the regions adjoining the C–C–A ends. J. Biol. Chem. *251*, 5124.
8. Khorana H.G. (1960) Synthesis of nucleotides, nucleotide coenzymes and polynucleotides. Fed. Proc. *19*, 931.

SUBJECT INDEX

Actin, interaction with phallotoxins 104–106
F-Actin 105, 106, 109
G-Actin 105
Ph-Actin 106
α-Actinin 110
Acyl xanthates, photolysis 26
Adenine arabinoside, antiviral activity 70, 75
S-Adenosyl-homocysteine 114
 analogs 114–118
 antitransforming activity 118, 119
S-Adenosyl-methionine 113
 reactions 115
 derivatives 117, 124
Alanine tRNA 197, 198
 function 198
 gene, synthesis 200
 processing 201
 structure 197
Alcohols, secondary, deoxygenation 32, 34
Aldosterone acetate 24
Aldosterone, 17a-hydroxy- 24
Alliin 171, 183
Alliinase 171, 175, 177, 179, 180
5-Allyl-2'-deoxyuridine, antiviral activity 70
5-Allyloxy-2'-deoxyuridine, antiviral activity 70
Amanitine 139
Amatoxins 97
Amino acid α-decarboxylases 178, 185
5-Amino-2'-deoxyuridine, antiviral activity 69
5'-Amino-5-iodo-2',5'-dideoxyuridine and analogs, antiherpes activity 60
 phosphorylation by thymidine kinase 60
4-Aminoisoxolid-3-one, see cycloserine
Antamanide 139–141
Antibiotic A-23187 145–148
Antibiotic X-537A 145–148
Antimycins 91
Antiviral agents, synthesis, properties 59–82
Aplasmomycin 87, 88, 93, 96
 antibiotic activity 93
 constitution and configuration 93
 X-ray analysis 88, 93

Arabinosyl nucleosides, antiviral activity 70, 75
ATPase 105, 109

Bacteriophage T4 220, 222
Bacteriophage λ 220–222
Bacteriophage φ 80 psu$_{III}$ 209
 r- and l-strand 209, 210
5-Benzyloxy-2'-deoxyuridine, antiviral activity 70
Bilayers 130
Biomembranes, ion transport 129–150
 structure
Bond angles, calculation 6–9
 in chromium carbonyls 10
 in cobalt carbonyls 10–13
 in iron carbonyls 11–14
 in manganese carbonyls 13–15
Boric acid, degradation product of boromycin 87
 in products of plant and animal origin 87
 addition to culture medium 87
Boromycin 87–89, 92–94
 alkaline hydrolysis 88
 antibiotic activity 87
 boric acid as degradation product 87
 boron-containing organic compound 87
 cavity 94, 96
 configuration of boron atom 92
 constitution and configuration 89
 empirical formula 88
 isolation 87
 lipophilicity 87
 spectroscopic data 88
 tertiary structure 96
 D-valine as degradation product, 88, 92, 94
 yield increase in biosynthesis 87
5-Bromo-2'-deoxyuridine, antiviral activity 70

(−)-Camphanic acid, absolute configuration 49
(+)-Camphanylchloride 49
Carbon atom, electronic structure 1
Carbonyl cyanide m-chlorophenylhydrazone 133

Carbonyl cyanide p-trifluoromethoxyphenylhydrazone 133
5-Carboxamidomethoxy-2'-deoxyuridine, antiviral activity 70
Cardiolipin 148
Catecholamines 147
5-Chloro-2'-deoxyuridine, antiviral activity 70
Chlorothiocyanogen, reaction with uridine and analogs 62
3,5-Cholestadiene 27
Cholesterol acetate dichloride, fluorination 35
Cholesterol thiobenzoate, photolysis 27
α-(L)-Cladinose, 39, 41
Cluster compounds, electronic structure 19
Cobalt atom, electronic structure 3
Cobalt complexes, structure 3, 4
Colletodiol 91
Colletoketol 91
Colletol 91
Corticosteroids, synthesis 30
Corticosterone acetate 24
   11-nitrite, photolysis 24
Crocetin, photosynthesis 23
Crown ethers, see macrocyclic polyethers
Cryptands 144
β-Cyanoalanine synthase 171, 172, 175, 178
5-Cyano-2'-deoxyuridine, synthesis 63
   antiviral activity 69
5-Cyanopyrimidine nucleosides, synthesis 62
   antiviral activity 69
5-Cyanouracil arabinoside, antiviral activity 75
5-Cyanouridine, synthesis 63
Cyclic dithiohemiacetals 43–45
   optically active, 49, 50
      configurational stability 49
α-Cycloglutamates 173
Cyclopeptides 141
Cycloserines, D and L 172–178, 180, 181, 184, 185
   as alanine antagonists 172, 173
   inhibition of lyases 174, 175
      mechanism 176, 177

   as pseudosubstrates of lyases 173, 181
   formation of O-substituted hydroxyl amines 173
γ-Cystathionase 171, 175, 177, 179, 180, 185
Cystathionine-β-synthase 171, 174
Cysteine, biosynthesis and metabolism 167
   as lyases inhibitor and substrate 171, 181, 182
Cysteine lyase 171, 174
Cytosine arabinoside, antiviral activity 70, 75

Demethylphalloin, structure 101
3-Deoxy-glucose, synthesis 32
5'-Deoxy-5'-S-isobutyl-adenosine 115
   analogs 116–117, 122
   activity, antimalarial 121
      antimitogenic 121
      antitransforming 118
   effect on methylation, nucleic acids 123
      proteins 124
   effect on amino acid incorporation 124
   inhibitory activity 115
   mechanism of action 122
   toxicity and metabolism 124
5'-Deoxy-5'-S-(2-methylpropyl)-adenosine 118
   antiviral activity 119
5-Deoxy-sugars, synthesis 33, 34
6-Deoxy-sugars, synthesis 33, 34
2'-Deoxyuridine 64, 67, 68, 73
   5-halogenosubstituted derivatives, antiviral and antimetabolic properties 68
   nitration by nitronium tetrafluoroborate 64
   incorporation into DNA 73
Des-alanine-1-phalloidin 100
Desboro-desvalino-boromycin (DBDVB) 88
   acetaldehyde as degradation product 88, 92
   acid $C_{13}H_{22}O_4$ as degradation product 88, 92
   constitution 91
   neutral compound $C_{18}H_{32}O_5$ as degradation product 88

β-(D)-Desosamine 39, 41
Desvalino-boromycin (DVB) 87
  alkaline hydrolysis product of boromycin 88
  antibiotic activity 93
  caesium salt 89, 94
  cavity 94, 96
  constitution 89
  enantiomer-selectivity 87, 95
  ionophore 87, 94, 95
  lithium salt 94
  NMR studies 94
  D-phenylglycine derivatives 95
  sodium salt 88, 94
  rubidium salt 89, 94
  tertiary structure 90, 92
  tetramethyl ammonium salt 94, 95
  X-ray analysis 89
Dethiophalloidin, CD curves 99
  structure 101
Diacetoneglucofuranoside 3-xanthate, reduction 32
1,2-Dicarbodecachlorododecaborane 133
β-Dicarbonyl compounds, enolisation, acylation and reduction 55
Dicoumarol 133
Dicyclohexylcarbodiimide 193, 195
Dienones, photolysis 23
Dihydroxy-L-leucin 98, 101, 102
Dinactin 142
2,4-Dinitrophenol 133
Dixanthates, reduction to olefines 32, 33
DNA biosynthesis inhibition by 5-substituted 2'-deoxyuridines 72
  principles for synthesis 197–200
  recombinant 223
  subdivision into segments 204
  Watson–Crick structure 191, 192
DNAase I, pancreatic 105
cis-Dithiadecalin system 42, 43, 47, 53, 55
  desulphurisation 42, 43, 53
DOPA decarboxylase 175

α,β-Elimination, lyase catalyzed 167–169, 172, 181, 183, 187
  mechanism 169, 183
β,γ-Elimination, lyase catalyzed 168, 187

Endonuclease EcoR$_1$ 211, 219, 220, 222
  recognition sequence 211, 212
Endonuclease P 219
Enneacovalency, transition metals 2, 4, 5, 15–18
Enniatins, 138, 139
Erythromycin 39, 40, 41, 47, 49
  conformation 39, 40
  nuclear moiety, synthesis 42
  stereochemistry 40, 41
  total synthesis 39–58
  key intermediates 51, 53
  X-ray studies 48, 49
Erythronolide B 42
5-Ethyl-2'-deoxyuridine, antiviral activity 60
N-Ethylmaleimide, action on β-replacing lyases 169, 184, 185

Ferrichromes 144
Fluorination 35
5-Fluoro-2'-deoxyuridne, antiviral activity 70
14-Fluoroprogesterone, synthesis 35

Glutamate α-decarboxylase 173, 179
5-Halogeno-deoxynucleosides, antiviral activity 68

Homocysteine, action on β-replacing lyases 182
  as cosubstrate of lyases 188
  as inhibitor of eliminating lyases 181
Homoserine as γ-cystathionase substrate 179
Hybrid bond orbitals, quantum mechanical treatment 6–10
  theory 6, 9, 15, 17, 19, 20
Hydrazones, oxydation 28, 29
Hydroxy acids, cyclisation to macrocyclic lactones 42
5-Hydroxy-2'-deoxyuridine, O-alkylation 65
16-Hydroxyheptadecanoic acids 91
5-Hydroxymethyl-2'-deoxyuridine, antiviral activity 70
allo-L-Hydroxyproline 98, 103
13-Hydroxytridecanoic acid, substituted 41

5-Hydroxyuracil arabinoside, antiviral activity 75
Hypofluorites, as fluorination agent 35

Imino-ethers, reduction, hydrolysis 25, 26
Interferon 114
γ-Iodoamids, cyclisation 25, 26
5-Iodo-2'-deoxyuridine, antiviral activity 70
   mechanism of action 60
Ion channels 130, 131, 150
Ionophores, complexes with metal ions 131, 132, 136–148
   conformation 136–147
   intrinsic 148–150
   mechanism of action 130–150
   specific for calcium 141, 145–148
'Ionophoroprotein' 148
Isophotosantonic acid lactone 22

Kephalin 148
Ketens, reactions 23
Ketophalloidin, structure 101

γ-Lactones, synthesis 25, 26
Lactose operon 209
Lanosterol, acetate 21
   transformation 22
Lasalocide, see antibiotic X-537A
Lead tetraacetate-iodine reagent 25
Limonin, oxydation 28
Liposomes 130
LIV protein 151–153
LS protein 151–153
Lumisantonin 22
Lyases, pyridoxalphosphate dependent 167–168
   catalytic mechanism 167–170, 172, 186
   cosubstrates 168, 170, 171
   β-eliminating 170, 171, 179–181
   inhibition 173–175, 179, 181
   isotopic exchange in catalysis 168, 180–182, 187
   β-replacing 170–172, 178–181, 183, 185–187
      distinction 182–186
      replacing agents 182

β-specific 170, 173, 181, 182
   substrates 171, 182

Macrobicyclic polyethers, see cryptands
Macrocyclic polyethers 142–144
Macrodiolide 90, 91
   biogenesis 91
Macrolides 39, 40
Macrotetrolides 141–142
5-Mercaptopyrimidine nucleosides, synthesis 62
Methylases, see transmethylases
Methyl-γ-benzyloxybutyrate 43, 44
   formyl derivative of 43, 44
Methymycin 42
Mitochondria 131, 147–150

Nortius, see macrotetrolides
Neopentyl aldehyde, transformation 28, 29
Nigericin 145, 146
5-para-Nitrobenzyloxy-2'-deoxyuridine, antiviral activity 70
5-Nitro-2'-deoxyuridine, synthesis and antiviral activity 64
Nitronium tetrafluoroborate, nitration of nucleosides 64
5-Nitropyrimidine nucleosides, synthesis 64
   antiviral activity 68, 69
5-Nitrouracil arabinoside, antiviral activity 75
5-Nitrouracil, preparation 64
5-Nitrouridine, synthesis 64
Nonactin 142
Norphalloin, structure 110

Oestrone, transformation 26
   18-hydroxy 26
Olefines, synthesis 32, 33
Oligonucleotides, chemical synthesis 193–197
   condensing agents 193–195
   enzymatic joining 204–208, 213, 223
   protecting groups 193, 194
      lipophilic (hydrophobic) 194, 196, 223
Oxidative phosphorylation, uncouplers 133

Penicillamine 181, 182
Pentachlorophenol 133
Phallacidin, affinity to rabbit muscle actin 98
  structure 98
  toxicity 98
Phallacin, affinity to rabbit muscle actin 98
  structure 98
  toxicity 98
Phallisacin, affinity to rabbit muscle actin 98
  structure 98
  toxicity 98
Phallisin, affinity to rabbit muscle actin 98
  structure 98
  toxicity 98
Phalloidine 139
  affinity to F-actin 103, 104
  affinity to rabbit muscle actin 98
  analogs, affinity to rabbit muscle actin 103, 104
    CD-curves 99, 100
    structure 101, 102
    toxicity 103, 104
  complex with actin 103, 104
  CD-curves 99, 100
  interaction with actin 104, 106
  possible molecular mechanisms of toxicity 109–111
  structure 98
  toxicity 98, 107, 108
Phalloin, affinity to rabbit muscle actin 98
  analogs, affinity to rabbit muscle actin 104
    structure 104
    toxicity 104
  structure 98
  toxicity 98
Phallotoxins 97, 98
α-Phenylethylamine 144
Phosphatidylcholine 148
Phosphatidylethanolamine 148
Phosphatidylserine 148
Plastoquinones 134
Polynucleotide kinase 198, 199
Polynucleotide ligase 198, 199, 205–208, 220, 222

Promoter 201–203, 210
  length of 211
Propargylglycine 184, 185
Prophalloin, affinity to rabbit muscle actin 98
  structure 98
  toxicity 98
5-Propynyloxy-2′-deoxyuridine, antiviral activity 70
5-Propynyloxyuracil arabinoside, antiviral activity 75
Prostaglandins 149, 150
Protonophores 133
Pyrenophorol 91
Pyrenophorin 91
Pyridoxylidene amino acrylates 168, 169
Pyridoxal phosphate, in lyases 167–178, 180, 181
  aldimine and ketimine forms 168, 170, 173, 176, 180, 181, 187
Pyridoxamine 173, 178, 179, 185
α-Pyridoxyl-β-aminooxyalanine 179
^{3}H-Pyridoxylcycloserine 178, 179
Pyrimidine nucleosides 5-substituted, see 5-substituted pyrimidine nucleosides

Quadruple bond, in chromium compounds 18
  in molybdenum compounds 19
  in rhenium compounds 17, 19

Racemases 173
β-Replacement, lyase catalyzed 167, 169, 170, 178, 187
  mechanism 169, 172, 182, 183
γ-Replacement, lyase catalyzed 168, 187
RNA polymerase B (II) 97

SAH, see S-adenosyl-homocysteine
SAM, see S-adenosyl-methionine
Santonin 21
  photolysis 22
Secophalloidin, CD curves 99
  lactone 102
  structure 101
Seleno-esters, reduction 34
Serine dehydratase 171, 175, 177, 179, 180

Serine sulfhydrase 171, 172, 174, 178
SIBA, see 5'-deoxy-5'-S-isobutyl-adenosine
iso-SIBA, see 5'-deoxy-S'-S-(2-methylpropyl)-adenosine
Sphingomyelin 148
Steroids, epimerisation at C_{13} 31
  17-ketoximes, acetylation 31
  nitrones, reactions 29
Streptomyces antibioticus, boromycin producent 87
5-Substituted 2'-deoxyuridines, antiviral and antimetabolic properties 68−75
  effect on DNA synthesis 72
5-Substituted pyrimidine nucleosides, synthesis and antiviral activity 59−82
Sugar, thione benzoates, reduction 32, 33
  thione carbonates, methylation 33, 34
    reduction 33, 34
  xanthates, reduction 32

4,5,6,7-Tetrachlorosalicylanilide 133
Tetrahydro-γ-thiapyrone 43, 44
  dimethyl ketal 43, 44
Terminator 201, 210
Thiazanes, as inhibitors of lyases 181
Thiazolidines, as inhibitors of lyases 181
Thiobenzoic acid 27
5-Thiocyano-2'-deoxyuridine, antiviral activity 70
5-Thiocyanopyrimidine nucleosides, synthesis 62
  biological activity 69
5-Thiocyanouracil arabinoside, antiviral activity 75
Thiono-esters 34
D-Threonine 98, 103
Thymine arabinoside, antiviral activity 70, 75
Transaminase, Ala:Glu 174
  Asp:Glu 174
  D-amino acids, 173
  mechanism of inactivation, 176, 177
Transition metals, carbonyls, bond angles 9−15
  resonance structure 15, 16
  covalency 2, 3
  enneacovalency 2, 4, 5, 15−18

enneacovalent radii 15−17
octahedral complexes, bond lenghts 17
  resonance structures 15−17
  stability 15
oxidation state 2
Transmethylases, inhibitors 113−114
  natural 114
  substrate analogs 114
  synthetic 114−118
Tributyl tin hydride 33, 34
Trifluoroacetamide 35
5-Trifluoromethyl-2'-deoxyuridine, antiviral activity 70
2-Trifluoromethyl-4,5,6,7-tetrachlorobenzimidazole 133
Triisopropylbenzene sulfonyl chloride 194, 195
Trinactin 142
Tropomyosin 110
Tryptathionine 98
Tryptophanase 168, 171, 175, 183
β-Tyrosinase 183
Tyrosine tRNA, precursor 201, 202
  gene 204, 208
  processing 202, 214, 217, 219
  structural gene, synthesis 203
  suppressor gene 200, 201, 216
    biological activity 217−222
    cloning 220−222
    nucleotide sequence 201
    promoter 211, 212
      elements of symmetry 212
      sequence and synthesis 208−216
    region adjoining CCA end, elements of symmetry 214
      sequence and synthesis 208−216
    synthesis 216, 217
    transcription 202, 217, 219
    transcriptional control elements 202

Ubiquinones 133
Uracil and analogs, nitration by nitronium tetrafluoroborate 64
Uracil arabinoside, antiviral activity 70, 75
Uridine, nitration by nitronium tetrafluoroborate 64
  and analogs, reaction with chlorothiocyanate 62

D-Valine, degradation product of boromycin 88, 92, 94
Valinomycin 136–137
  analogs 137
Vermiculin 91
Vijlsmeir reagent 34

5-Vinyl-2'-deoxyuridine, antiviral activity 70
Vinyl iodides 28

Xanthate radicals 26